T0243076

CAMBRIDGE LIBRARY COLLECTION

Books of enduring scholarly value

Technology

The focus of this series is engineering, broadly construed. It covers technological innovation from a range of periods and cultures, but centres on the technological achievements of the industrial era in the West, particularly in the nineteenth century, as understood by their contemporaries. Infrastructure is one major focus, covering the building of railways and canals, bridges and tunnels, land drainage, the laying of submarine cables, and the construction of docks and lighthouses. Other key topics include developments in industrial and manufacturing fields such as mining technology, the production of iron and steel, the use of steam power, and chemical processes such as photography and textile dyes.

The Construction of Roads, Paths and Sea Defences

Professionally dedicated to civil engineering and architecture, Frank Latham became Penzance Borough Engineer and Surveyor in 1899, served as a captain in the Royal Engineers during the First World War, and designed the famous Penzance lido for the silver jubilee of George V. For a paper on the town's sea defences, the Civil Engineers of Ireland awarded him a medal in 1905. He had considerable practical experience in his field when he published this work in 1903. Originally written as a series of articles for a municipal engineering journal, the text was revised and presented to readers as an illustrated work of reference. In addition to his thorough explanation of the methods of road and sea-defence construction, Latham includes an appendix on the prices of the materials required. Contemporary advertisements for relevant companies and their various products further enhance the book's interest to historians of technology.

Cambridge University Press has long been a pioneer in the reissuing of out-of-print titles from its own backlist, producing digital reprints of books that are still sought after by scholars and students but could not be reprinted economically using traditional technology. The Cambridge Library Collection extends this activity to a wider range of books which are still of importance to researchers and professionals, either for the source material they contain, or as landmarks in the history of their academic discipline.

Drawing from the world-renowned collections in the Cambridge University Library and other partner libraries, and guided by the advice of experts in each subject area, Cambridge University Press is using state-of-the-art scanning machines in its own Printing House to capture the content of each book selected for inclusion. The files are processed to give a consistently clear, crisp image, and the books finished to the high quality standard for which the Press is recognised around the world. The latest print-on-demand technology ensures that the books will remain available indefinitely, and that orders for single or multiple copies can quickly be supplied.

The Cambridge Library Collection brings back to life books of enduring scholarly value (including out-of-copyright works originally issued by other publishers) across a wide range of disciplines in the humanities and social sciences and in science and technology.

The Construction of Roads, Paths and Sea Defences

With Portions Relating to Private Street Repairs, Specification Clauses, Prices for Estimating, and Engineer's Replies to Queries

Frank Latham

CAMBRIDGE
UNIVERSITY PRESS

University Printing House, Cambridge, CB2 8BS, United Kingdom

Cambridge University Press is part of the University of Cambridge.
It furthers the University's mission by disseminating knowledge in the pursuit of education, learning and research at the highest international levels of excellence.

www.cambridge.org
Information on this title: www.cambridge.org/9781108072090

This edition first published 1903
This digitally printed version 2014

ISBN 978-1-108-07209-0 Paperback

Inside Front Cover.

THE

CONSTRUCTION OF ROADS, PATHS, AND SEA DEFENCES.

WITH PORTIONS RELATING TO

PRIVATE STREET REPAIRS, SPECIFICATION CLAUSES, PRICES FOR ESTIMATING, & ENGINEER'S REPLIES TO QUERIES.

BY

FRANK LATHAM, C.E.,

Member of the Institution of Civil Engineers, Ireland; Member of the Society of Engineers; Member of the Sanitary Institute of Great Britain; Member of the British Association of Water-works Engineers; Fellow of the Institute of Architects and Surveyors.

Borough Engineer and Surveyor, Penzance.

Author of "The Sanitation of Domestic Buildings," "The Dispersion of Noxious Gases in Sewers and Drains," &c.

London:

THE SANITARY PUBLISHING COMPANY, LIMITED

5, FETTER LANE, E.C.

1903.

London:

PRINTED BY GEO. REVEIRS, GRAYSTOKE PLACE, FETTER LANE, E.C.

PREFACE.

SEEING the great advances that have been made in recent years in all branches of road construction and the many engineering works connected therewith, and being aware of the increasing demand for additional and improved roads, which must be regarded as a first necessity for the welfare and comfort of every civilised and developing country, the author penned in his spare moments (with the assistance of many leading engineers, to whom he is much indebted for valuable data), a lengthy series of articles for the columns of the "Sanitary Record and Journal of Municipal Engineering," with the desire of producing in an interesting form a useful record of the more modern methods of construction, materials, and maintenance, adopted in roads and paths, in the hope that they would be read from week to week as they were published, and that some amount of benefit to the public would result therefrom.

Although the leisure time of a busy official is limited, yet no pains were spared in obtaining useful data for these articles, a fact which seems to be recognised, several appreciative letters having been received from readers expressing their hope that the writings would be reproduced in book form, a desire which is now acceded to, but not without revision of all matter gone before; the whole of the articles have been carefully revised and a quantity of additional matter added, which should enhance the value of the volume as a book of reference.

The author is conscious that, whilst much care has been given to the accuracy and fulness of the information contained in this work, imperfections exist in some of its pages, treating more particularly with collateral branches, that could only be dealt with to advantage in separate volumes.

The author hopes that the information supplied in the following pages will be found useful to the student and others interested in the subject, which is a matter of practical everyday and growing importance to the public.

F. L.

Public Buildings, Penzance,
January, 1903.

THE CONSTRUCTION OF ROADS, PATHS AND SEA DEFENCES.

CHAPTER I.

INTRODUCTION.

THERE are few questions which have such a practical bearing on the progress of a nation, the advancement of civilisation, the furtherance of knowledge, and the numerous advantages to be derived from easy intercourse between all parts of a country, as that of the effectual construction and maintenance of roads.

It is a matter of much importance in these days of competitive industry and increasing needs of commerce, bringing in their wake the ever-growing demand for improved facilities of exchange and transport, and the more extensive adoption of modern modes of locomotion, that all roads should be constructed in the most efficient matter, so as to best conform to the requirements of the times; thus their durability of construction and those conditions necessary for safety and ease in travelling should be carefully considered, and the best course adopted, thereby reducing as much as possible the amount of wear and tear to horses and vehicles, and at the same time contributing to the comfort of those whose business causes them to be "on the road."

The economies derived from a system of good roads over even secondary ones must be considerable, but when compared in these days with the bad roads of a century or more ago, the saving to the public must be enormous, besides abolishing an inconvenience which would be absolutely intolerable to the high standard of excellence expected by the public at the present day.

Sir Henry Parnell says:—"The measures necessary to be taken for affording the means of travelling with rapidity and safety, and of transporting goods at low rates of carriage, form an essential part of the domestic economy of every people. The making of roads, in point of fact, is fundamentally essential to bring about the first change that every rude country must undergo in emerging from a condition of poverty and barbarism. It is, therefore, one of the most important duties of every Government to take care that such laws be enacted, and such means provided, as are requisite for the making and maintaining of well-constructed roads into and throughout every portion of the territory under its authority."

In populous districts, large towns, and health resorts, it is of paramount importance, apart from the desirability of good roads for traffic, that all roads and back courts should be maintained in a proper state of repair, from a sanitary point of view. A residential district with bad and neglected roads and courts cannot be regarded as healthy. Cleanly roads in proximity to dwellings play an important part in the sanitary education of the inhabitants, and no dwelling can be regarded as perfectly sanitary, especially in crowded localities, when abutting on neglected and, perhaps, narrow, sunless, and inefficiently-drained roads.

Since the introduction of railways there is practically greater demand for good roads in and about towns than was previously the case, the fact that railways have set in motion the whole of the people and commerce, calling for the provision of the best facilities for transportation to and from these main centres and arteries of the country.

It is by means of the railways that so many thousands of persons flock to the seaside towns and other health resorts at various seasons of the year, calling forth in no small degree the construction of beautiful roads, promenades, and drives, sometimes entailing much engineering skill and considerable outlay of money.

The railway has, by its rapid conveyance of commerce and passengers from one locality to another, placed the remotest parts of the country within easy touch with the great centres, and has been the means of opening out many delightful and hitherto unfrequented districts, all in their turn requiring improved means of intercommunication.

As the earlier history of the advancement of road making in England is interesting, it is as well that, before entering on to the present methods of construction, an outline of the past should be briefly sketched.

The Carthaginians have the credit for the invention of paved roads. The Romans, who were in touch with the Carthaginians from about 500 years B.C., became well instructed in the art of road making, and in the year 312 B.C., Appius Claudius partly constructed the most noble of Roman roads, named after him the Via Appia, or Appian Way. Fig. 1 shows the method adopted by the Romans in the construction of their principal roads.

The interest taken in the construction of roads by the Romans was exceedingly great, so much so that under Augustus and Julius Cæsar all the principal cities and towns were made to communicate with Rome by paved roads, which work was followed by numerous similarly constructed roads communicating throughout Europe and continuing through neighbouring continents and

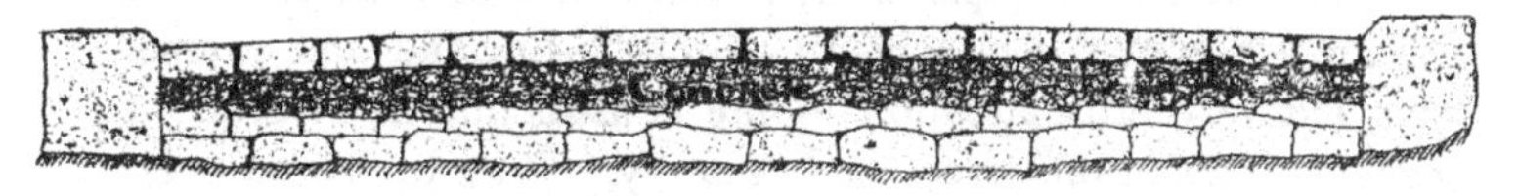

Roman Road

Fig. 1.

islands, the direct lines of the great roads being maintained from shore to shore.

Mr. Tredgold says:—"The Roman roads ran nearly in direct lines; natural obstructions were removed or overcome by the effort of labour or art, whether they consisted of marshes, lakes, rivers, or mountains. In flat districts the middle part of the road was raised into a terrace. In mountainous districts the roads were alternately cut through mountains and raised above the valleys, so as to preserve either a level line or a uniform inclination. They founded the road on piles where the ground was not solid, and raised it by strong walls or by arches and piers where it was necessary to gain elevation."

Pinkerton's Geography contains the following remarks on the Roman roads of England:—*The Roads of England*, "which may still be traced in various ramifications, present a lasting monument of the justice of their conceptions, the extent of their views, and the utility of their power. A grand trunk, as it may be called, passed from the south to the north, and another to the west, with branches in almost every direction that general convenience and expedition could require. What is called Watling-street led from Richborough in Kent, the ancient Rutupiæ, north-east through London to Chester. The Ermine-street passed from London to Lincoln, thence to Carlisle, and into Scotland. The Foss way is supposed to have led from Bath and the Western regions north-east till it joined the Ermine-street. The last celebrated road was Ikeneld, or Ikneld, supposed to have extended from near Norwich southward into Dorsetshire."

Upon the Roman power being broken up, their successors allowed the roads to fall into a bad state of repair, in fact cared little or not at all for them, so that in course of time these magnificent roads were practically ruined by absolute neglect and contempt.

It was not until comparatively recent date that successful road making was practised by modern nations. England and France gave but very little attention to the construction and repair of roads until about the seventeenth century. Prior to this time the roads of this country were allowed to remain in the most disreputable condition, and in some places were almost impassable. As recently as 1770 a publication by Mr. Arthur Young, describing his tour in the North of England, gives some very striking accounts of the state of the roads at that time. In describing some of the turnpike roads, he explains that the road from Wigan was so bad that he measured ruts in its surface 4ft. in depth, and that the road's surface was mended merely by "tumbling in some loose stones, which serve no other purpose than jolting a carriage in the most intolerable manner." The road to Newcastle appears to have been even worse than the Wigan one. Mr. Young says: "A more dreadful road cannot be imagined," and that he was obliged to hire two men at one place to prevent his chaise from overturning. He says:—"Let me persuade all travellers to avoid this terrible country, which must either dislocate their bones with broken pavements or bury them in muddy sand."

Mr. Thomas Codrington, in his work on "The Maintenance of Macadamised Roads," says: — "On the ordinary highways, improvement was hindered by the system of statute duty, and by parish management under a person chosen yearly to serve the office of surveyor of highways. Everyone who kept a team of horses was liable to be called upon to do six days' team work, and those who did not keep horses paid money instead. The parish surveyor generally had no special knowledge of road repairing, and the team labour and other work were seldom well applied."

Under this system the responsibility of maintaining roads near towns or between towns would come very heavy upon those few persons who were expected to provide the funds for carrying out the necessary repairs.

The first attempt to improve the highways was by the introduction of the turnpike system.

The first turnpike road was established by law in 1653, by means of which a toll was imposed upon all but pedestrians who passed through the gate. This toll was then used for purposes of road-making. It appears, however, that the people were prejudiced against the system of toll gates, and that for many years the turnpikes were few.

At the end of George the Second's reign the system was put into more general operation throughout the country.

It was, therefore, the end of the 18th century before any marked improvement was made in the defective state of the roads.

The following extract from an essay on the "Construction of Roads, dated 1813," by Mr. Edgeworth, will show the defective condition of roads that existed even in the early part of the 19th century. Mr. Edgeworth says:—"In many parts of this country, and especially near London, the roads are in a shameful condition; and the pavement of London is utterly unworthy of a great metropolis."

Early in the 19th century rapid changes began to take place in road making, largely due to the labours of Metcalf, Telford, Macadam, and others.

Mr. Macadam, in his "Remarks on the Present System of Road-making," says the then existing practice of making a road in England and Scotland was "to dig a trench below the surface of the ground adjoining, and in this trench to deposit a quantity of large stones, after this a second quantity of stones, broken smaller, generally to about 7 lb. or 8 lb. weight; these previous beds of stone are called the bottoming of the road, and are of various thicknesses according to the caprice of the maker, and generally in proportion to the sum of money placed at his disposal. On some new roads made in Scotland in the summer of 1819 the thickness exceeded 3ft. That which is properly called the road is then placed on the bottoming by putting large quantities of broken stone or gravel, generally 1ft. or 1ft. 6in. thick, at once upon it, and from the careless way in which it is done the road is as open as a sieve to receive the water which is retained in the trench, whence the road is liable to give way in all changes of the weather."

CHAPTER II.

THE NECESSITY OF GOOD ROADS FOR TRACTION.

SIR ISAAC NEWTON, writes Parnell, "has laid it down as a general principle of science, that a body, when once set in motion, will continue to move uniformly forward in a straight line by its momentum, until it be stopped by the action of some external force. This proposition is admitted and adopted by all natural philosophers as being perfectly true, and, therefore, in order to apply it to roads, it is necessary to inquire what kind of external forces act in a manner to diminish and destroy the momentum of carriages passing over them."

Sir Henry Parnell remarks, "As a carriage for conveying goods or passengers when put in action becomes a moving body, in the language of science, the question to be examined and decided is, how a carriage when once propelled can be kept moving onwards with the least possible quantity of labour to horses, or of force of traction."

Each of these great philosophers admits that some more or less controllable influences are constantly at variance with the principle of science known as the "first law of motion," and that the causes thus offering resistance should be ameliorated to as great a degree as possible. In this respect due regard must also be given to the necessary requirements for the comfort and safety of those who produce the propelling force to such traction, as well as keeping the object in view of reducing the force of traction.

Resistance to rolling is principally due to four causes, viz., surface protuberances and irregularities, surface and wheel friction, surface grades, and wind pressure.

Surface protuberances and irregularities.—When a wheel comes in collision with any hard road obstruction it becomes necessary for a horizontal force to be exerted at the axle to raise such weight carried by the wheel to a height equal to that of the obstacle to be passed over. In comparing a road of considerable protuberances and irregularities with one of smooth surface, the comparative loss of energy in drawing a heavy load must be a matter of no little moment. These irregularities are further responsible for resistance caused by the shocks, which are greater when a vehicle is drawn at a high speed than at a low one.

On ordinary hard roads the resistance attributable to the actual raising of the weight over trifling obstructions is, perhaps, less to be taken into account than that reponsible for the concussions which take place as the load is drawn over roughly paved surfaces, and in which, as M. Morin determined in his experiments, the resistance to traction is increasing with the speed.

In the case of the want of uniformity in the road surface the resistance may readily be arrived at by the application of the following formula :—

Let P = the force required.
W = the weight, for example taken at 300 lb.
R = the radius of wheel " 30in.
O = the obstacle " " 2in.

$$\text{Then} \quad P = W \frac{\sqrt{R^2 - (R - O)^2}}{R - O}$$

$$\therefore \quad P = 300 \frac{\sqrt{30^2 - (30 - 2)^2}}{30 - 2}$$

Springs to vehicles have a tendency to diminish the resistance offered by road inequalities at high speeds. Mr. Codrington says regarding this : "The effect of springs in reducing draught is that they enable the wheels to rise and fall over inequalities over the road, while the load on them moves forward without being sensibly raised. The more perfect the elasticity of the springs in a vertical direction the greater is the reduction of draught, but any elasticity in the direction of the traction tends to increase the draught."

Surface and wheel friction.—This form of resistance varies according to the material used in making the road, whether it be hard or soft ; also, to some extent, according to the size of the wheel. On soft roads the wheel will, to some extent, penetrate the surface, so that when in motion it will be obstructed by a force in front of it of either grit, dust, or by the weight of the vehicle causing an indentation sunk in the material itself. This, consequently, has the constant effect of causing the wheel to climb an irregularity equal to the depression so formed. The amount of penetration is less for large wheels than for small, but it is sufficient for general purposes, and for wheels of ordinary sizes, to calculate the resistance thus caused by penetration by the weight multiplied by a ratio to be determined on by the nature o the road surface, whether it be hard, soft, or medium. The following tables are the results of experiments by Sir John Macneil (made with an instrument he invented for the purpose) as to the resistance to traction of various descriptions of roads.

A wagon weighing 21 cwt. was used on different kinds of roads with the following results :—

	lb.
On well-made pavement the draught is	33
On broken stone surface, or old flint road	65
On gravel road...	147
On broken stone road, upon rough pavement foundation ...	46
On broken stone surface, upon a foundation of concrete formed of Parker's cement and gravel	46

The following empirical formulæ are given by Sir John Macneil for calculating the resistance to traction on level roads for (i.) a stage wagon; (ii.) a stage coach :—

P = force required.
W = weight of vehicle in lb.
w = weight of load in lb.
V = Velocity feet per sec.
C = A constant number depending upon the nature of the surface over which the carriage is drawn.

Then $$P = \frac{W + w}{93} + \frac{W}{40} + C V \quad \ldots \quad \ldots \quad \text{(i.)}$$

and $$P = \frac{W + w}{100} + \frac{W}{40} + C V \quad \ldots \quad \ldots \quad \text{(ii.)}$$

The value of C is as follows :—

On timber surface ... ... $C = 2$
On paved surface ... ... $C = 2$
On a well-made broken stone road ... $C = 5$

On a well-made broken stone road, covered with dust... ... ... $C = 8$
On a well-made broken stone road, wet and muddy ... ... $C = 10$
On a gravel or flint road, in a dry, clean state ... ... $C = 13$
On a gravel or flint road in a wet and muddy state ... ... $C = 32$

Example :—To find the force necessary to move a stage coach at a velocity of 10ft. per second, the coach weighing 1500 lb. and carrying a load of 700 lb., and the surface of the road being well-made of broken-stone, but wet and muddy.

Then, according to (ii.).

$$\frac{1500 + 700}{100} + \frac{700}{40} + (10 \times 10)$$

Grades.—Surface grades bring into account the additional force of gravity.

The tractive resistance due to the grade is that force which is necessary to support a wheel on an incline and to prevent it from running down, under the pressure of its load + the effects of friction.

Therefore the total tractive resistance to be arrived at is the sum of a given load hauled over a level road + the extra force required to draw the same load up to a graded road.

The following diagram (Fig. 2) is to represent a wheel C loaded with a weight W.

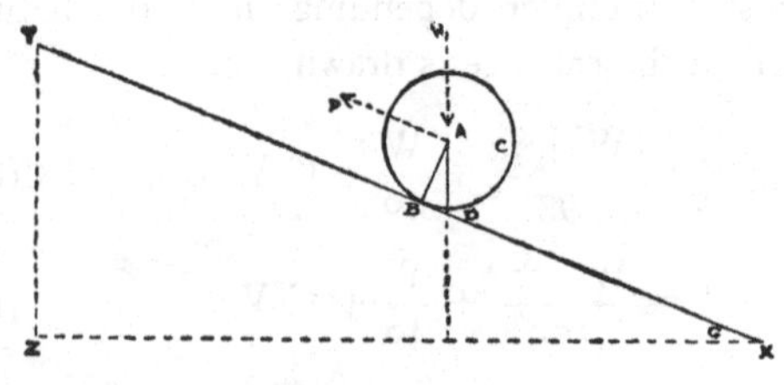

Fig. 2.

The wheel is in contact with the inclined plane X Y at B, and P is the force required to haul the load up the incline, and acting parallel to the road.

The wheel is maintained in position by these forces, viz., W, which at all times acts in a vertical direction. By the force P, which is the applied power parallel to the inclined plane, and by the pressure caused by the weight of the vehicle on the road's surface at B, and which is effected by a perpendicular from the point A with the inclined plane X Y.

Draw the vertical base X Z and the horizontal line Y Z, then the relative magnitude of the three forces at P, W and D may be determined.

Because A D and Y Z are parallel and are cut by the line X Y, then the angles A D B and X Y Z are equal, and the angle A B D is equal to the angle X Z Y, being both right angles, therefore the remaining angles B A D and Y X Z are equal, and the triangles X Y Z and A B D are similar.

Now, as the sides of the triangle A B D are proportional to the three forces by which the vehicle is sustained, so also are the three sides of the similar triangle X Y Z, viz.: X Y is proportional to W, Y Z to D, and X Z to the pressure B.

$$i.e.\ W : XY : : D : YZ,$$

$$\text{and } W : XY : : B : XZ.$$

Therefore if to X Z such a value be allotted that the vertical rise of the road is exactly 1ft., then :—

$$D = \frac{W}{XY} = \frac{W}{\sqrt{AG^2 + 1}} = W.\ \sin G,$$

$$\text{and } B = \frac{W.\ XZ}{XY} = \frac{W.\ XZ}{\sqrt{XZ^2 + 1}} = W.\ \cos.\ G.$$

In order to ascertain the force required to move a vehicle upon an inclined road (neglecting the effects of friction), divide the weight of the loaded vehicle by the length of the inclined surface, which affords a rise of 1ft., and the quotient will represent the force required.

Again, in order to ascertain the force with which such vehicle bears upon the surface of the same road, the horizontal length of the road is to be multiplied by the weight of the loaded vehicle and the product divided by the inclined length of same, and the result will be the pressure required.

The pressure of a wheel on an inclined road is slightly less than the actual weight, but the difference is very slight on grades of normal inclination.

To ascertain the resistance to the traction of a vehicle being drawn up or down an incline, first determine the resistance on a level road of such load, the surface to be taken as the same as that of the incline. To this result add or subtract, according to whether the vehicle is travelling up or down hill, the force requisite to sustain the same weight on the incline; then the sum of resistance on a level road and the force requisite on the incline will express the force required for ascending the incline, and the difference of the two results will be the force required for descending the incline.

When the grade force is greater than the surface resistance, it may be necessary to apply extra force to check the propelling force.

Mr. D. K. Clark, in the "Construction of Roads and Streets" (Law and Clark), says: "Rolling or circumferential resistance of wheels is equal to the product of the load by the third of the semichord (of the submerged arc of the wheel), divided by the radius of the wheel." He further remarks: "This question is no doubt applicable, with a sufficient degree of accuracy, for any real needs for calculating the resistance of gravel, loose stones, soft earth or clay." He then deduces as follows: "The circumferential or rolling resistance of wheels to traction on a level road is inversely proportional to the cube root of the diameter," and "to reduce the rolling resistance of a wheel to one-half, for instance, the diameter must be enlarged to eight times the primary diameter." M. Dupuit deduced that on macadamised roads of good condition, and on uniform surfaces, "the resistance to traction is directly proportional to the pressure. It is independent of the width of the tire. It is inversely as the square root of the diameter. It is independent of the speed."

M. Dupuit admits that on paved surfaces which give rise to constant concussion the resistance increases with the speed.

Mr. Thomas Codrington says: "The resistance due to gravity on inclined roads must, of course, be allowed for in all considerations of draught. It is very nearly equal to the gross load divided by the rate of gradient; thus, on a gradient of 1 in 20 the increase of draught due to gravity will be $\frac{1}{20}$th of the gross weight of the vehicle and its load."

Tractive power.—The tractive power of horses is influenced by still another important condition in addition to those stated, and that is the foothold afforded by the surface. The load drawn by a horse, although not generally affected on the level, must be regulated according to the grade and surface of the greatest incline it is to be drawn up.

Mr. Macneill, in his "Notes" on the "Holyhead Road," remarks to the following effect:—That when the road is not horizontal, the force of gravity is a great impediment to the draught of carriages, and considerably limits the effect which would otherwise be produced by a horse in drawing a load. If it were not for the hills that are usually met with on turnpike roads, one horse would do as much work as four; for it is well known that the force of draught must be commensurate to the steepness of hills. This being the case, it is important that the greatest facilities should be given to render the ascent as easy as possible by paving such grades with suitable materials that would afford the best foothold.

The difference between the maximum load that a horse is able to draw on smooth asphalt, compared with that on a good macadamised surface, is about 50 per cent. in favour of asphalt on the level; and the average foothold that a horse obtains on all grades up to 5 per cent. on good macadam, as compared with smooth asphalt, is about 75 per cent. in favour of the former.

The tractive power of horses varies very considerably, according to the work and description of pavements that they have been accustomed to. It is, therefore, difficult to arrive at anything but an approximate idea on this subject.

The force that a horse may exert may be taken inversely to the rate at which it travels. An average horse that works ten hours a day at a speed of 250ft. per minute on a horizontal road should be able to pull a draught of 75 lb. This would be equivalent to lb. × feet × time (in minutes).

$$\therefore 75 \times 250 \times 600 = 11{,}250{,}000 \text{ foot-pounds,}$$

represent one day's horse work.

The following will show the tractive force as effected by the amount of resistance offered by the road surface. From experiments made by Sir J. W. Bazalgette: Gross load, four tons; speed from four to six miles per hour:—

	Tractive force on the level.
Macadamised surface	40·7 to 42·29 lb. per ton.
Asphalt	39·0 to 39·32 lb. per ton.
Wood	33·62 to 36·63 lb. per ton.
Granite setts	26·2 to 27·0 lb. per ton.

Comparative statement given by Mr. H. P. Boulnois in his "Municipal and Sanitary Engineers' Hand-book," showing the traction upon level roads formed of different surfaces; asphalt being taken as the standard of excellence:—

Asphalted roadway	1·0 —
Paved roadway, dry and in good order	1·5 to 2·0
" " in fair order	2·0 to 2·5
" " but covered with mud	2·0 to 2·7
Macadamised roadway, dry and in good order	2·5 to 3·0
" " in a wet state	3·3 —
" " in fair order	4·5 —
" " but covered with mud	5·5 —
" " with the stones loose	5·0 to 8·2

CHAPTER III

LINE OF ROAD.

BEARING the question of gradients in mind, the next matter for consideration is the location of a road which would provide the shortest route for the main portion of the traffic.

In planning new roads it is often found necessary to deviate from a direct line, in order to meet certain existing roads or to provide for the traffic to some village, or owing to the desirability of taking a pleasanter route. So that it is seldom that a road can be laid out in such a direct line as could be desired from an economic point of view.

Apart from these, there are other influences which are too often allowed to affect the line of a road, and these require a larger amount of serious attention than is generally given to them. These causes consist generally of the expense of constructing bridges over rivers, or of raising a road by means of embankments above low and flat lands subject to inundation during high floods.

Sir Henry Parnell remarks on this point: "The peculiar circumstances of a river may render it necessary to deviate from a direct line in laying out a road. A difficulty may arise from the breadth of the river, requiring a bridge of extraordinary dimensions, or from the land for a considerable distance on the side of the river being subject to be covered with water to the depth of several feet in floods. In these cases it may appear, upon accurately calculating and balancing the relative inconvenience and expense of endeavouring to keep a straight line and of taking a circuitous route, that upon principles of security, convenience, and expense, the circuitous course will be the best.

"In general, rivers have been allowed to divert the direct line of a road too readily. There has been too much timidity about incurring the expense of new bridges, and about making embankments over flat land to raise the roads above the level of high floods.

"These apprehensions would frequently be laid aside if proper opinions were formed of the advantages that arise from making roads, in the first instance, in the shortest directions and in the most perfect manner. If a mile, half a mile, or even a quarter of a mile of road be saved by expending even several thousand pounds, the good done extends to posterity, and the saving in annual repairs and horse labour that will be the result will before long pay off the original cost of the improvement."

But little consideration should be required to convince all concerned that where practicable the route selected for a road should be the one with the least hills and gradients along its course, provided that the length of such road would not be unduly increased in so doing.

The importance of this matter is being felt more and more as the modern methods of locomotion come into use.

A steep gradient to ascend on an outward journey means one to descend on the return. The risks incurred in descending steep hills are sometimes considerable, resulting at times in serious accidents, and the additional energy required in ascending similar inclines often necessitates the employment of extra horse labour; at the best of times they are fatiguing and entail much loss of time.

A further important consideration is, that the amount of repair and attention that steep gradients require is much above that of level roads, chiefly due to the use of skids on the wheels of vehicles, and to erosion of the metal caused by water scour on the surface.

With modern roads, with their smooth surface, there is frequently a danger of a horse slipping and falling down in descending steep inclines, and there is also a difficulty for the horse to gain the desired foothold to enable it to draw a weight up such gradients.

The steepest gradient that should prevail in the line of a new road—excepting where, owing to circumstances, it is necessary to depart from the rule—is that known as the "ruling gradient."

This "ruling gradient" is that inclination to be determined which will require but a moderate additional force in ascending, and no difficulty or danger in descending, and will present no practical impediment to any ordinary traffic.

Mr. Thomas Codrington considers that the ruling gradient on a macadamised road should be 1 in 30. He says: "On a level macadamised road in ordinary repair the force which the horse has to put forth to draw a load may be taken as $\frac{1}{30}$th of the load. But in going uphill the horse has also to lift the load, and the additional force to be exerted on this account is very nearly equal to the load drawn divided by the rate of gradient. Thus, on a gradient of 1 in 30 the force spent in lifting is $\frac{1}{30}$th of the load, and in ascending a horse has to exert twice the force required to draw the load on a level. In descending, on the other hand, on such a gradient, the vehicle, when once started, would just move of itself without pressing on the horse. A horse can, without difficulty, exert twice his usual force for a time, and can therefore ascend gradients of 1 in 30 on a macadamised surface without sensible diminution of speed, and can trot freely down them."

Messrs. Law and Clark say: "Sir John Macneil, in 1836, maintained that no road was perfect unless its gradients were equal to or less than 1 in 40. In thus limiting the ruling gradient to 1 in 40, he justifies the assertion by the much greater outlay for repairs on roads of steeper gradients. For instance, he adduces as a fact not generally known, that if a road has no greater inclinations than 1 in 40, there is 20 per cent. less cost for maintenance than for a road having an inclination of 1 in 20. The additional cost is due, not only to the injury by the action of the horses' feet on the steeper incline, but also to the greater wear of the road by the more frequent necessity for sledging or braking the wheels of vehicles in descending the steeper portions."

Sir Henry Parnell considers 1 in 35 to be the ruling gradient, and remarks as follows:—"An inclination of 1 in 35 is found by experience to be just such an inclination as admits of horses being driven in a stage-coach with perfect safety when descending in as fast a trot as they can go; because, in such a case, the coachman can preserve his command over them, and guide and stop them as he pleases. A practical illustration that this rate of inclination is not too great may be seen on a part of the Holyhead Road, lately made by the Parliamentary Commissioners on the north of the city of Coventry, where the inclinations are at this rate, and are found to present no difficulty to fast driving, either in ascend ing or descending. For this reason it may be taken as a general rule, in laying out a line of new road, never, if possible, to have a greater inclination than that of 1 in 35. Particular circumstances may, no doubt, occur to require a deviation from this rule; but nothing except a clear case that the circuit to be made to gain the prescribed rate would be so great as to require more horse labour in drawing over it than in ascending a greater inclination, should be allowed to have any weight in favour of departing from this general rule. On any rate of inclination greater than 1 in 35 the labour of horses in ascending hills is very much increased. The experiments detailed in the Seventh Report of the Parliamentary Commissioners of the Holyhead Road, made by a newly-invented machine for measuring the force of traction or power required to draw carriages over different roads, fully establishes this fact."

The following is a table contained in the appendix of Sir Henry Parnell's work of the general results of experiments made with a stage-coach on the same description of road; but on different rates of inclination, and with different rates or velocity, on the Holyhead Road :—

Rates of inclination.	Rates of travelling.	Force required.
1 in 20	6 miles per hour	268 lb.
1 in 26	6 " "	213 lb.
1 in 30	6 " "	165 lb.
1 in 40	6 " "	160 lb.
1 in 600	6 " "	111 lb.
1 in 20	8 " "	296 lb.
1 in 26	8 " "	219 lb.
1 in 30	8 " "	196 lb.
1 in 40	8 " "	166 lb.
1 in 600	8 " "	120 lb.
1 in 20	10 " "	318 lb.
1 in 26	10 " "	225 lb.
1 in 30	10 " "	200 lb.
1 in 40	10 " "	172 lb.
1 in 600	10 " "	128 lb

Where long, steep inclines cannot be avoided, it is desirable to divide them up into lengths, by introducing at intervals short lengths of road of less gradient, which will serve to relieve the heavy strain from horses when drawing loads up the hill.

In some instances where new roads are necessary, but owing to the elevated contour of the land, and circumstances surrounding the question, a steep incline cannot be avoided, then it is sometimes advisable to lessen the incline by cutting through the summit.

The incline of a hill may in some instances be lessened by making up the road at the foot with either the material removed from the summit, or by means of other suitable substance. This road should be banked up, with slopes on each side, which, if necessary, may be further supported by retaining walls.

This procedure can often be adopted with considerable advantage in overcoming deep valleys. It is far from an uncommon occurrence to meet with a steep descent to be almost immediately followed by an ascent; in such cases it would be an improvement if the valleys between were more or less filled up. This operation not only saves a lot of tedious up and down-hill work, but also shortens the distance. Fig. 3.

In locating a line for a new road, a preliminary reconaissance should be made. It is necessary, in so doing, that the principal characteristics of the country through which a new road is to be carried should be carefully noted.

This work is much simplified if a reliable map of the district is obtained, but should such maps not be procurable, then the work of reconnoitering will be considerably increased.

Upon the map—which should accompany the engineer when making his examination of the country — the points which appear to afford the most suitable lines can be noted, and fixed with sufficient accuracy to enable sketches to be made previously to a full survey with levels; after which the route can be finally laid out.

Contour maps, showing the principal elevations of the country, are sometimes available, and in such cases the work to be done is much facilitated, as the greater portion of the planning, after the preliminary examination, can be done on the map.

In considering the lines selected, note should be made as to the available material for embankments, where these appear to be necessary; the nature of the material to be excavated where such work is required, and in doing this the particular geological formations, over which it is proposed that the lines should pass, should be ascertained. The position of quarries from which stone and gravel could be obtained should be noted, together with any

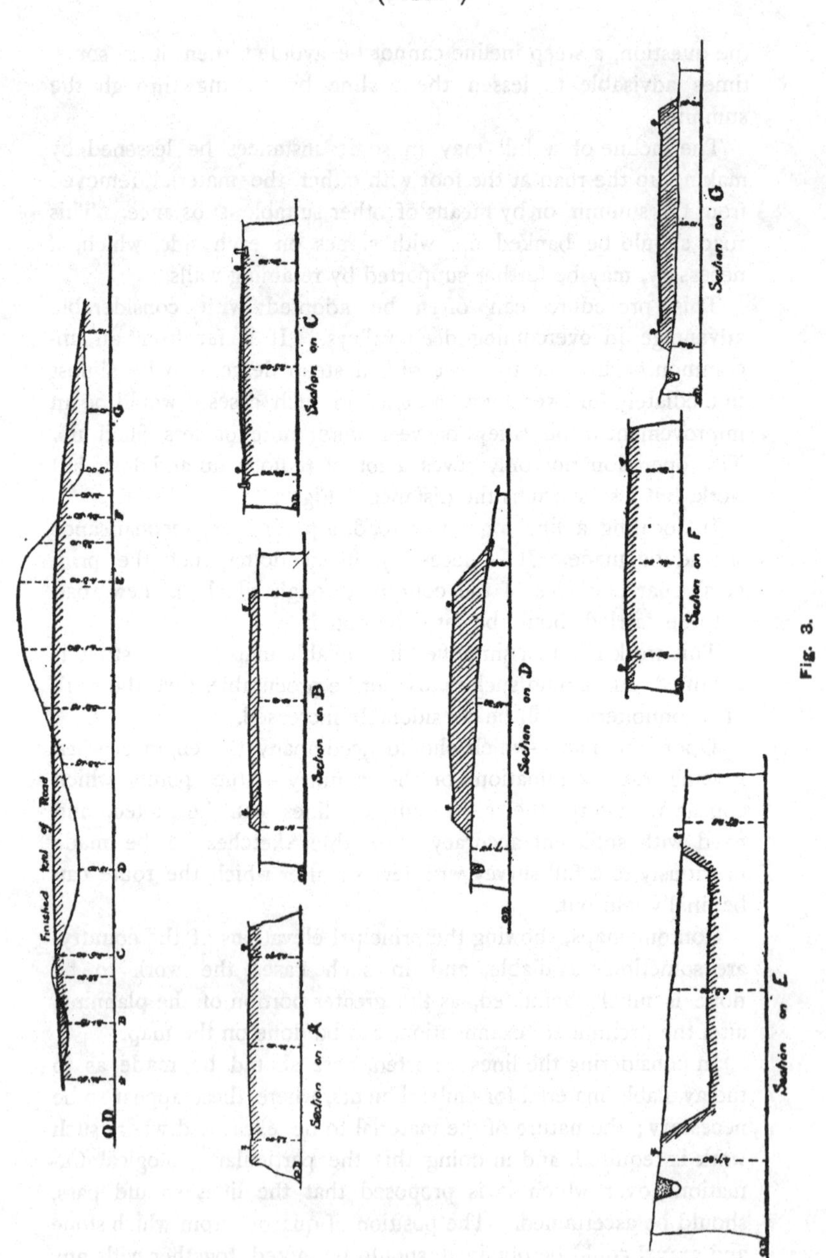
O.D.
Finished level of Road
Section on A
Section on B
Section on C
Section on D
Section on E
Section on F
Section on G

Fig. 3.

general circumstances and conditions that must be in favour of or against any particular line that might be selected.

It is desirable, in cases where a turning or curve takes place on an incline, for the safety and convenience of traffic, this deviation from the direct line should be of as large a radius as possible, and the incline at such points should be reduced.

It is sometimes found desirable to alter the line of an existing road for the purpose of reduction of distance, or improvement in gradients. This question, whatever advantages may be derived from the diversion, is one of much importance; not only the convenience of those who use the road should be considered, but the possible injury to property by the alteration must be borne in mind, or otherwise cases of compensation may arise. The advantage of such diversions to the public generally should carry much weight, but care should be taken that as little injury as possible is caused to neighbouring property.

Upon the line of a road being finally determined, the next business is to stake out the route. This is generally done by driving a line of wooden stakes into the ground all along the course of the proposed road. These stakes are placed at about 100ft. apart for straight roads and closer together at curves, and are so fixed as to define the centre line of the work; levels are then taken along the line at each stake, from which a longitudinal section is drawn. Cross levels are also taken at intervals where considered necessary. Upon these longitudinal and cross sections the finished levels are indicated.

Width of Road.—The width of the roadways of towns up to recent years has been too often insufficient for the conduct of the traffic, and in many cases considerable outlay has been necessary to remedy these congested conditions, in purchasing and removing property for street-widening purposes. This is an experience that nearly every town has had at one time or another. In country roads situate between towns of importance, and thereby subjected to much traffic, the width of such roads should not be less than 36ft., and should include a 6ft. path on one side, but in ordinary country roads not so situate the width is not of so much importance, as it is seldom that more than two vehicles are required to pass each other at a time. A width sufficient to meet the requirements of the traffic in each direction is all that is needed, and the cost of an unnecessary width of roadway to construct and maintain is avoided. Such roadways therefore require to be only about 18ft. in width. Although the roadway need not be of greater width, yet it is most desirable that the distance between hedge and hedge should be of the full width of 30ft. to 36ft. This may be accomplished by leaving a strip of grass land at each side through which footpaths can be made and sub-drains carried. By so arranging the width, the road is exposed to the full benefit of the sun and wind, which is essential to obtaining a dry and good wearing surface.

On this subject Mr. Macadam, in answer to the question, "What width would you in general recommend for laying out a turnpike-road?" remarks :—"That must depend upon the situation. Near great towns roads, of course, ought to be wider than further in the country. Roads near great towns ought not to be less than 30ft. or 40ft. wide; but at a distance from great towns it would be a waste of land to make them so wide." (Evidence before a Committee of the House of Commons, 1819.)

Road Boundary Walls, &c.—Hedges and close-boarded fences should not be placed at the sides of country roads unless they are kept low enough or at sufficient distance away as not to obstruct the action of the wind and sun. The free exposure of country roads to both the drying effects of the wind and sun are most essential, in order to maintain the surface in good condition for any length of time. Attention to this is important, not so much from the fact of the convenience and comfort afforded by a dry road, but that the effect of wet has a great tendency to weaken the road and cause it to wear more quickly and cut under the weight of the traffic.

Trees in streets also have an injurious effect on roads, and sometimes are the cause of a road being kept wet for a considerable time by the dripping of water from their leaves.

From another point of view, however, it is desirable to have trees in roads, as they afford shade from the blaze of the sun, which in some streets in summer-time is almost unbearable to horses as well as to pedestrians. Trees are, further, a great attraction, and beautify the perspective of thoroughfares. In the country, where trees are planted on each side of the road, special attention should be given to see that no hedges or walls are allowed to exist along the line to obstruct the free action of the wind under the trees.

In selecting trees for street planting the description of tree must be considered in conjunction with the locality and conditions to which they will be subjected. Trees of light foliage and small growth are more suited for this purpose than the more dense and larger variety. Where planted in pavements it is advisable to

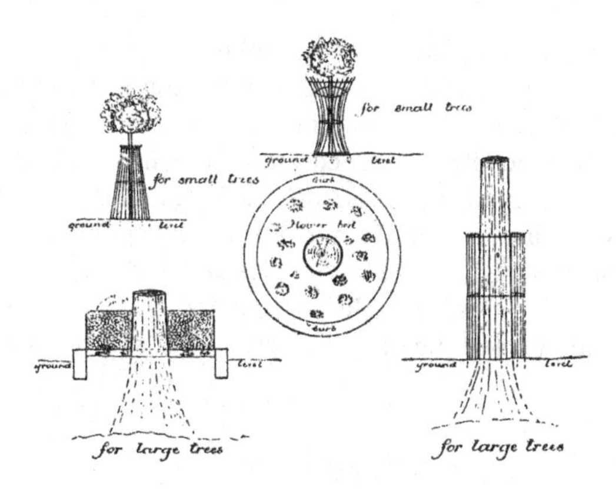

Fig. 4.

place an iron grating over a portion of the earth surrounding the tree, so as to admit air and a small quantity of water to the roots; the earth over the roots is in some towns allowed to remain exposed and frequently used as a flower bed. It is, further, advisable that all young trees growing in streets should be protected by a tree guard, as illustrated—Fig. 4—to prevent the trunk from being injured by horses or mischievous persons.

CHAPTER IV.

EMBANKMENTS.

IN raising a road by means of embankments, it is essential that much care should be exercised in executing the work, in order to produce a well-consolidated and firm construction. Indifference to this course generally means constant settlement and trouble, and not infrequently is the cause of serious sideslips.

It is obvious in the case of roads of this description that unless the work be executed in the most substantial manner the extra expense that will necessarily follow in maintaining the surface in good condition must be considerable. Those who have had

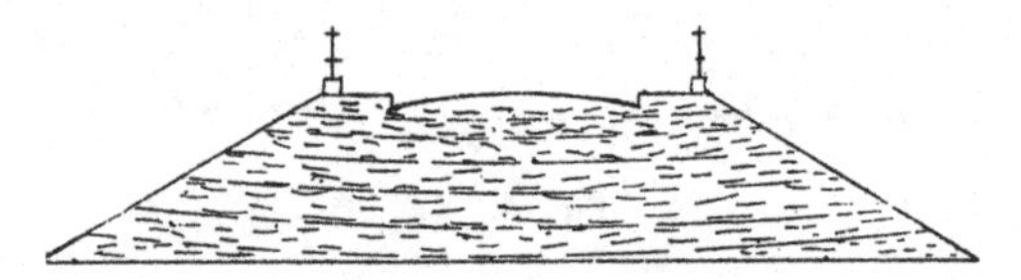

Fig. 5.

charge of roads, after completion, with weak foundations, know too well the difficulty of maintaining their surfaces for any length of time in a state of reasonable repair.

These embankments are usually constructed by means of dump-wagons; these wagons are generally made to run on temporary rails over the newly-made surface, and upon arriving at the "tip," their contents are discharged over the end from the top of the embankment. This is an expeditious method, but not so thoroughly to be relied upon as that recommended by Sir Henry Parnell. He advises that "in forming high embankments the earth should be laid on in concave courses . . . . in order to give firmness and stability to the work," and that courses "not exceeding 4ft. in thickness" should be adopted (Fig. 5).

When embankments are constructed on the tipping principle, the filling should be done from the sides towards the centre, which will cause the material to arrange itself in layers, with a dip from the sides inwards.

With regard to the slope necessary to be given to the side of an embankment or cutting, this should always be greater than the inclination that the earth naturally assumes, and which varies according to the nature of the soil, as will be observed from the following details given by Sir H. Parnell:—" In the London and plastic clay formation it will not be safe to make the slopes of embankments or cuttings that exceed 4ft. high with a steeper slope than 3ft. horizontal for 1ft. perpendicular. In cuttings in chalk or chalk marl the slopes will stand at one to one. In sandstone, if it be solid, hard, and uniform, the slopes will stand at a quarter to one, or nearly perpendicular.

" If a sandstone stratum alternate with one of clay or marl, it is difficult to say at what inclination the slopes will stand; this will, in fact, depend upon the inclination of the strata. If the line of the road is parallel to the line of the bearing of the strata, in such cases large masses of the stone become detached, and slip down over the smooth and glassy surface of the subjacent bed. There are many instances of slips in sandstone and marl strata under such circumstances as those now described, and here the slopes are as much as four to one. If the road is across such strata, or at right angles to the line of bearing, then the slopes may be made one and a-half to one; but if the strata lie horizontal, even though there should be thin layers of marl between the beds of stone, the slopes will stand at a quarter to one. But it will be necessary if the beds of marl exceed 12in. in thickness to face them with stone. " In the Oxford clay, which covers so great a portion of the midland counties of England, the slopes should not be less in any instance than two to one, and even in some parts of this formation they should be made three to one if the cuttings be deep. In all such cases, if there be any beds of gravel or sand found intermixed with the clay . . . drains should be cut along the top, and even in the sides of the cuttings; for if this precaution be not taken, the water, which will find its way into the gravel, will, by its hydrostatic pressure, force the body of clay down before it, and slips will take place even when the inclinations are as much as four to one; and when this occurs it is extremely difficult to re-establish them.

"In limestone strata, if they be solid, slopes will stand at a quarter to one; but in most cases limestone is found mixed with clay beds, and in such cases the slopes should be one and a-half or two to one. In the primitive strata, such as granite, slate, or gneiss, slopes will stand at a quarter to one."

Another method of constructing embankments is by spreading thin layers of material on at a time of from 1ft. to 1ft. 6in. in thickness, which are in their turn consolidated by ramming or rolling. This is a process most valuable in special cases requiring extra strength.

Embankments require some amount of attention as to drainage. The surface should be formed in the shape of a roadway, with a footpath on each side, and the surface water conducted into side channels, to be at intervals removed by drains, otherwise this water would run over the sides of the embankments and injure the

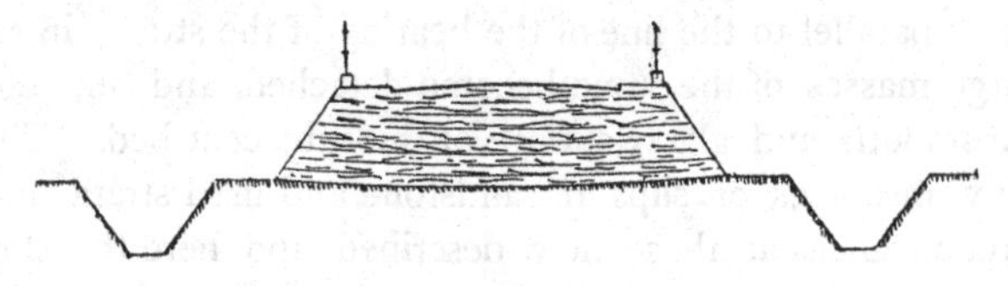

Fig. 6.

slopes. Embankments also require drainage at their base to prevent the water in times of rainfall or flood, to as great a degree as possible, from injuring the stability of the construction. This drainage may be effected, when the ground is firm, by ditches at each side of the embankment. The material excavated in so making these side ditches will generally be found useful in constructing the embankment (Fig. 5). At places where springs are crossed it is necessary that the water should be piped through into the side ditches.

It may be found necessary at times to carry embankments over marsh or bog land. In these cases, unless special care be taken, the foundation will give way under the weight placed upon it, and the embankment will in consequence sink into the soft ground. In order to overcome this difficulty a good deal of discretion on the part of the engineer is necessary, as the degree of difficulty will vary according to circumstances.

Several methods have been successfully adopted to meet these difficulties, and may be thus briefly described :—In some cases the first work to be taken in hand is the drainage of the land on each side of the line of the proposed new embankment. This is usually effected by means of open ditches—Fig. 6—which, as a rule, cause the land water to be removed from the site over which the embankment is to pass, and thus the foundation is rendered firmer. If this course fails to produce a sufficiently firm foundation, then a further expedient would be to excavate the ground for a foot or so along the line to the width of the base of the embankment, and to fill this trench in with suitable large stones or other bottoming (Fig. 7).

Embankments have been carried over treacherous ground by using broad wooden platforms, constructed of hurdles and boards, or sometimes faggots or dry peat used either separately or in con-

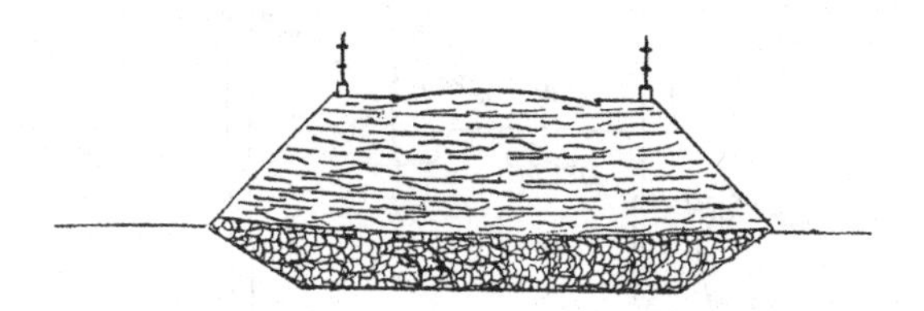

Fig. 7.

junction with the wooden platform. When these materials are employed, they are placed over the soft ground, and the embankment is completed to the height required by means of light materials such as peat. The surface is finally laid with macadam spread over a layer of faggots or hurdles. The whole construction, thus being as light as possible, forms a structure that practically floats on its soft marshy bed.

EMBANKMENTS ON "SIDELONG" GROUND.

The information given by Mr. D. H. Mahan in "A Treatise on Civil Engineering," with regard to this work, may be here quoted with advantage :—"When the axis of the roadway is laid out on the side slope of a hill, and the road surface is formed partly by excavating and partly by embanking out, the usual and most simple method is to extend out the embankment gradually along the whole line of excavation. This method is insecure, and no pains therefore should be spared to give the embankment a good footing on the natural surface upon which it rests, particularly at the foot of the slope. For this purpose the natural surface should be cut into steps, or offsets, and the foot of the slope secured by buttressing it against a low stone, or a small terrace of carefully rammed earth.

"In side-formings along a natural surface of great inclination, the method of construction just explained will not be sufficiently secure; sustaining walls must be substituted for the side slopes,

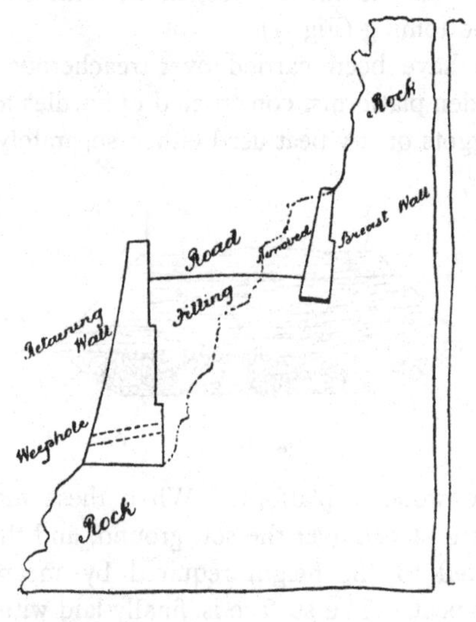

Fig. 8.

both of the excavations and embankments. These walls may be made simply of dry stone, when the stone can be procured in blocks of sufficient size, to render this kind of construction of sufficient stability to resist the pressure of the earth. But when the blocks of stone do not offer this security, they must be laid in mortar, and hydraulic mortar is the only kind which will form a safe construction. The wall which supplies the slope of the excavation should be carried up as high as the natural surface of the ground; the one that sustains the embankment should be built up to the surface of the roadway, and a parapet wall should be raised upon it, to secure vehicles from accidents in deviating from the line of the roadway. A road may be constructed partly in excavation and partly in embankment along a rocky ledge, by blasting the rock when the inclination of the natural surface is not greater than one perpendicular to the base; but with a greater inclination than this, the whole should be in excavation.

"There are examples of road construction, in localities like the last, supported on a framework consisting of horizontal pieces, which are firmly fixed at one end by being let into holes drilled in the rock, and are sustained at the other by an inclined strut

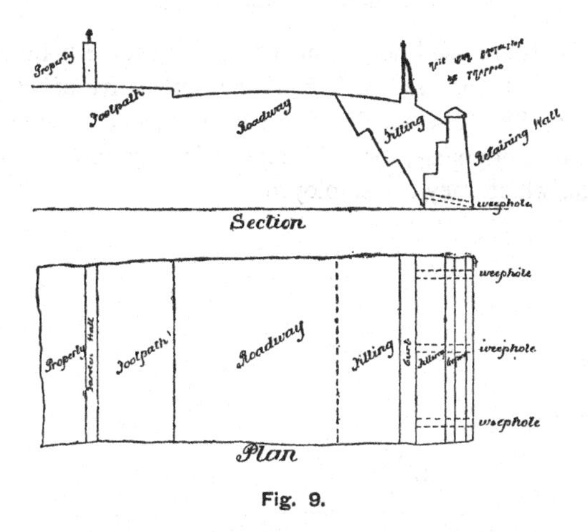

Fig. 9.

underneath, which rests against the rock in a shoulder formed to receive it."

RETAINING WALLS.

It is often desirable, and sometimes necessary, to adopt retaining walls in forming embanked roads along steep inclines or in special circumstances. These walls should be constructed of brick, stone, or concrete, and be of sufficient strength to resist the thrust of the earth behind them under all conditions. Sir Benjamin Baker considers that the thickness of retaining walls, in ground of an average character, should be equal to one-third of the height of the wall, measured from the top of the footings; and in cases where the backing and foundation are both favourable, then a wall one-quarter of the height so measured, and with a batter of 1in. or 2in. per foot on the face, would be sufficient. A retaining wall that is slightly curved in the batter, as in Fig. 8, is to be preferred to one with a straight batter, as it is better able to support the pressure to which it is subjected from behind. In building retaining walls in more or less weak ground, it is advisable that the weight of the wall should be distributed over as large an area as possible; the base must therefore be proportioned accordingly. Fig. 9 gives a plan and section of a road supported by a retaining wall.

BREAST WALLS.

Breast walls are used not so much to resist a force behind them, but to protect the natural earth from the destructive effects of the weather (Fig. 8). These walls, as a rule, need not be of any great thickness, but, of course, that will depend largely upon the conditions under which they are employed.

CHAPTER V.

THE construction of embanked roads along the foreshore, sides, or frontage of towns situate on the sea coast often calls for much consideration. Most seaside towns of enterprise sooner or later possess their promenades and marine drives. The building of roads along the foreshores of towns by gaining land from the sea, as is often the mode followed, is perhaps the most difficult branch of road engineering, and much skill and care are required in the operation. This work almost enters into a separate branch of engineering, and is much too large a subject to be dealt with to any extent in these pages. The methods usually adopted in embanking a road between the sea on the one side and the land on the other are of two characters, viz., embankments protected by walls and embankments without walls.

The weight of a wall or embankment necessary to resist the force of the sea and waves is an essential consideration. Sea walls should be constructed sufficiently substantial to oppose and resist the attacks of the heaviest seas that are likely to occur, without causing undue shock to the road supported by the walls, or in any way damaging the construction of the walls themselves. The safety of the promenade or road is bound up in this important factor. No hard-and-fast rule can be made to apply equally to all requirements around the coast as to the description, the weight, and other conditions necessary or desirable for the protection of walls exposed to the attacks of the waves. The force that the sea will exert upon any particular wall or embankment will vary at different parts of the coast, according to the amount of exposure and depth of water, and the extent of the stretch of sea or ocean before it.

For approximate calculations the following particulars may be useful:—The weight of sea water is 64·11 lb. per cubic foot, or 1·027 the weight of fresh water.

The effects of pressure and velocity of the wind on this water produces more or less momentum, and the water is thus agitated and driven with great force against the face of any structure opposing it. The force of the wind per square foot in a storm is 12·300 lb., and in a hurricane the force is 31·490 lb.

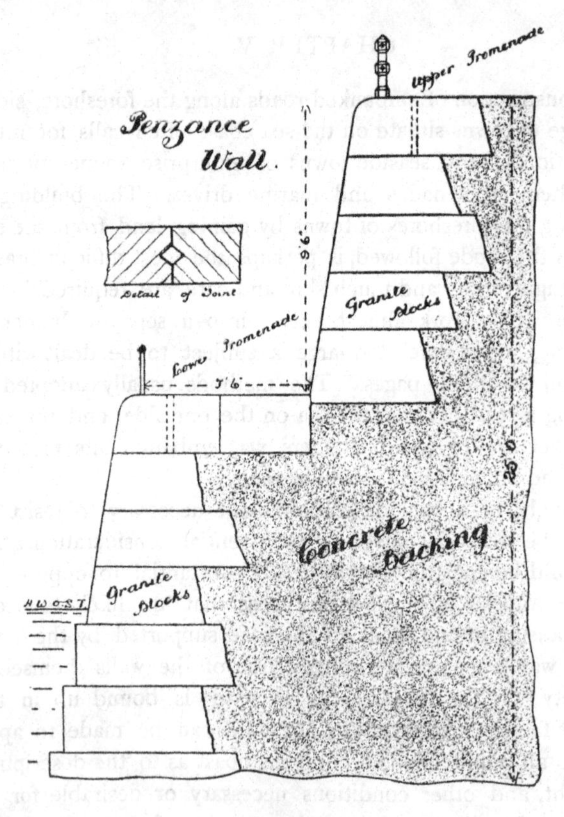

Fig. 10.

Coasts open to the Atlantic are subjected to not only the surface wave, but a great disturbance of the whole depth of the water. Whenever there is a storm in these waters, a ground swell is created, consisting of a continual oscillation backwards and forwards, augmented by the great mass of water, and forming a most destructive element to sea walls and foundations.

In deep water in the Atlantic the force of the waves and rollers has been calculated to be as much as 6083 lb., or nearly 3 tons to the square foot, and at Dunbar as high a force as 3½ tons has been registered. At Dunbar Harbour a force equal to 45·04 cwt. was registered on a dynamometer in 10¼ft. of water, the observation being taken 9ft. 6in. above bottom. In such circumstances it is far less costly in the long run to construct proper walls for sea defences than to lay out money on a wall that will sooner or later be demolished. The wall shown in Fig. 10 is one recently constructed by the Penzance Corporation in front of their beautifully situated and extensive promenade. This wall replaced an almost perpendicular one, which from time to time collapsed from the action of the waves. The projecting base of this wall tends to break the force of the sea, and minimises the quantity of water thrown on to the promenade, as compared with other structures. Walls in similar positions should be of sufficient strength to resist much greater pressure than the force of the sea, so that the impact will do the minimum injury to the backing or promenade. Other conditions besides exposure must be considered in determining the dimensions necessary for a wall, such as the quality of the materials employed in its construction, the form of profile of the wall, the nature of the foundation, and the description of backing.

MATERIALS.

The materials used in the construction of sea defence walls should be the best available; they should be strong and hard and of high specific gravity. Most of the stones employed in road-making are suitable for sea walls, such as granites, basalt, green-stones, gneiss, porphyries, and schists. Concrete consisting of high-class Portland cement and good clean hard stone and sand is often employed with much success. This is either made in blocks on the site of the works in a somewhat similar manner to the manufacture of concrete paving slabs, or the concrete is built up *in situ*. In the latter method the forming of the wall is effected by a casing of timber and boards, the boarding being carried up a little ahead of the concreting as the work proceeds, and being removed when the concrete wall is properly hardened. Most of the manufacturers of concrete paving slabs for footpaths and promenades also make blocks of the same kind of material suitable for sea walls. These blocks are made to any design, and are finished with perfect exposed surfaces, equal to the paving slabs. Blocks made of concrete, faced with harder stones of small dimensions, bonded and toothed into the concrete, make a splendid material for sea walls where there is much heavy sea. This method is being adopted in the new Dover Admiralty Pier works, by which very heavy and strong blocks are made with the use of a small quantity of granite. The author was much interested in the manufacture of these blocks whilst residing but a few miles from the works, and they appeared to him to be the

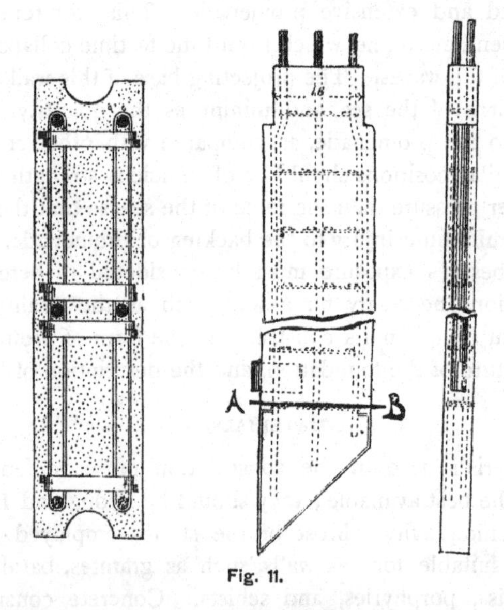

Fig. 11.

most excellent description for sea wall works, combining both strength and economy of cost. A system of patent concrete piles has recently been introduced into England from France, known as Hennebique's Patent Ferro Concrete. Several important quay and heavy structural works have already been successfully carried out in this country on this system.

Many of the drawbacks and difficulties of setting masonry under water, and of the risks in all sea works of losing work whilst in course of construction, by rough seas, are avoided by the application of Hennebique's patent system of concrete sheet piling. With this system, a good foundation can be obtained for a wall equally as well under water as on dry land, thus assuring a good foundation to any desired depth. The sheet piles are designed with a longitudinal and half cylindrical groove on each edge, and a projecting tongue is fixed on the bottom end of each pile—Fig. 11 —so that when one pile is in its place, the tongue of the next follows the groove of the previously driven pile, the grooves acting as a guide for the adjoining pile. When two or more piles have been driven, the grooves between them form a cylinder, into which water is forced to clear away any earthy or other substances, and subsequently each cylinder is filled up with grout cement concrete, which, when properly done, makes the joint firm and water-tight.

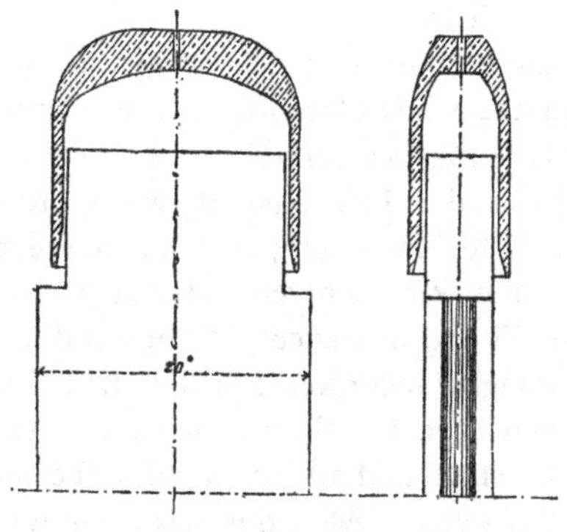

Fig. 12.

The piles are moulded with Portland cement concrete, internally strengthened by a system of iron or steel rods, interlaced and arranged, and of such thickness best suited to the nature of the work in which the piles are to be employed (Fig. 11). Each pile is shod with steel, and a steel helmet of suitable shape is placed over the head of the pile before driving (Fig. 12). The space between the helmet and head of pile is filled with sand or sawdust, and a timber dolly is interposed between the monkey and the pile head, as a precaution to prevent the impact damaging the pile.

Concrete used for sea walls should be made as follows:—The aggregate should be of small size, ranging from ½in. to 2½in. It is essential that the aggregate should be of different sizes, the same as in road stone for road-making, so that the smaller sizes may fill up the voids between the larger stones. Sharp hand-broken stone forms the best aggregate; round pebbles or smooth beach stones should not be used excepting for backing. To fill in the interstices between the smaller stones a certain quantity of clean sharp sand is necessary, and finally a sufficient quantity of cement is required to fill up the remaining voids. One cubic yard of good concrete will take about 1 cubic yard of sharp stone broken from ½in. to 2½in. gauge; ½ cubic yard of clean sharp sand to fill the interstices between the stones; ¼ cubic yard of finely ground Portland cement to fill the interstices in the sand. The method of mixing the concrete best suited for exposed sea walls is of much importance, for its strength much depends on this being properly attended to.

Concrete is sometimes made mechanically, but is usually mixed by hand, the materials being placed in their respective proportions on a platform, where they are turned whilst dry, and again whilst wet, till properly mixed. This method, when properly done, is quite good enough for most purposes, but a greater amount of care is desirable in making concrete for sea walls. Seeing that the cement is to fill the interstices of the sand, these materials require to be thoroughly incorporated; it is therefore best to well mix the cement and sand together in their respective proportions, first whilst in a dry state, and afterwards with the addition of the required quantity of water. After this has been done, the broken stone, after being damped with clean water, so that it should not soak up the moisture from the matrix, should be thoroughly mixed with the prepared cement and sand, and when ready for use the mixture should be free of excess of water.

The crushing resistance of concrete made in the foregoing manner, of the best materials, and allowed to become hard by exposure to the air, should be about 1000 lb. per square inch, and the transverse strength should be about 250 lb. per square inch.

The bedding surfaces of blocks for sea walls should be purposely roughed to make a good key for the jointing material, which will prevent the block from sliding or being moved when struck by the sea or any floating object. However strong the materials of which the wall is constructed may be, the wall will be rendered weak if the joints are improperly made. Each block should be properly bedded in cement mortar, and the blocks driven as closely together as possible, so that the joints on the face of the wall may have the appearance of being practically closed.

PROFILE OF WALL.

Walls of almost perpendicular form are the most unsuitable and expensive; walls so constructed offer too sudden an obstruction to the force of the sea, and therefore need to be of heavier masonry than walls built with a sloping, angular, or corrugated

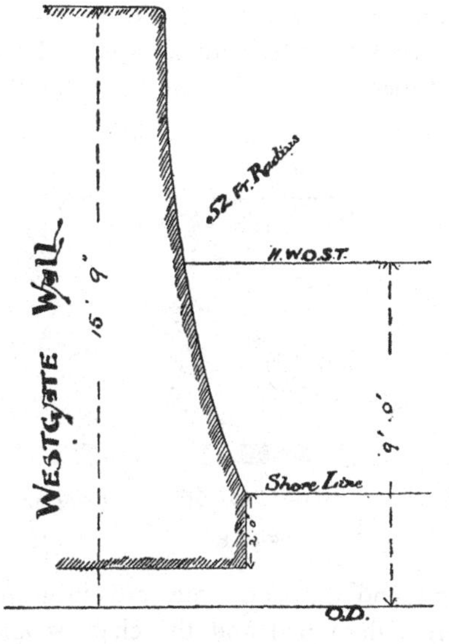

Fig. 13.

sectional profile. A greater amount of disturbance takes place at the foot of upright walls than in the case of walls with a batter. The author has had before his notice for some time a sea wall constructed of granite blocks; this wall was originally built with a slight batter, but is now falling forward, and is 1in. in 1ft. out of the perpendicular. The road supported by this wall settles down at times of heavy storms at sea. It appears that the waves strike the lower portion of the wall and cause it to shake or vibrate, which further causes a disturbance in the backing, setting up a creeping action between the outward vibrations of the wall and the filling supported by it, and causing the wall to be gradually pushed out at the top.

Promenade sea walls should be of such a profile that the waves should be gradually broken up with as little effect on the foundations as possible, so as to diminish scour. Many forms have been tried with this object in view, but few have been successful. At Westgate-on-Sea, near Marga e, a wall constructed of concrete of curved form—as shown in Fig. 13—was observed for some years by the author. This wall was mentioned by Mr. Richard F. Grantham, M.I.C.E., in his valuable paper read before the Society of Engineers in 1897. He says: "The chalk rock on the shore appeared hard, and was covered with seaweed. The base of the wall was built 2ft. deep below the shore level. But in two or

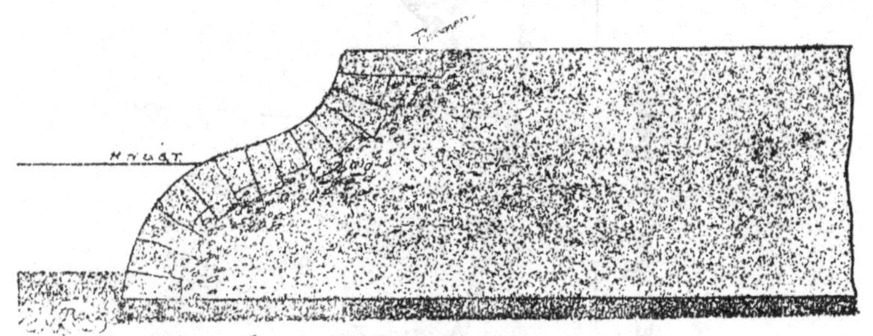

Western Promenade Wall margate

Fig. 14.

three years it was found that for some yards in length the foot of the wall had been undermined and the chalk scoured out about 9in. The remaining lengths of the foundations were built 4ft. below the shore level."

The minutes of the Association of Municipal and County Engineers contain an account of the Scarborough wall, which was constructed with a radius of 17ft. The shale at the foot of this wall was scooped out to a depth of 3ft. within a year of the work being completed. When the curved form of wall is adopted the foot should, in cases where the foundation is other than hard rock, be protected by means of an apron of masonry or concrete, continued for a few feet from the toe of the wall, which will better resist the erosive action of the sea than the natural formation.

The O.G. wall, constructed many years ago at Margate—shown in Fig. 14—has stood many gales, but was much damaged by an extraordinary sea of abnormal height in November, 1897. The damage it sustained on this occasion was greatly owing to inferior workmanship. This wall is built on a chalk foundation, and no trouble from undermining has ever been experienced; the portions

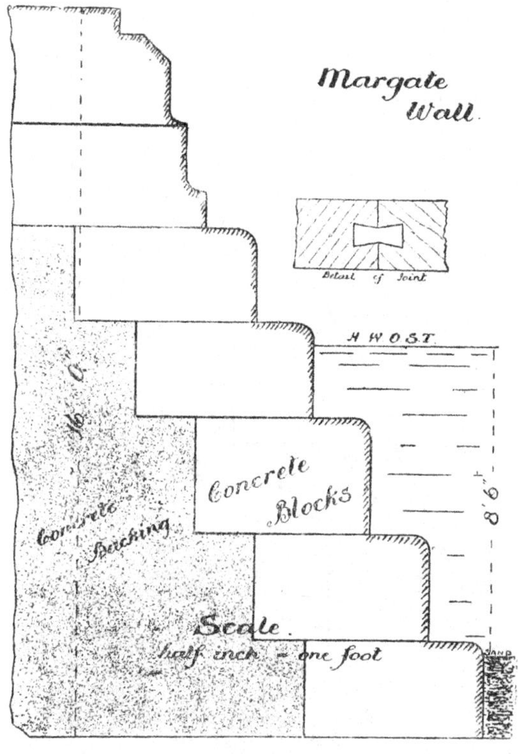

Fig. 15.

damaged by the great storm in 1897 were above the projecting base. The wall is built from 12in. to 18in. thick, merely backed up with rammed and grouted chalk, and has stood intact for nearly half a century. The unprotected chalk cliffs within a stone's throw of the wall were washed away 4ft. 6in. in some places during the year of the above-mentioned storm, whilst in other years it averaged about 2ft. per annum.

The Marine Drive wall, Margate, and a later wall, in the construction of which the author took part, were designed and constructed by Mr. Albert Latham, M.I.C.E. The form of these walls is shown in Fig. 15. The steps in this form of wall serve to break up the force of the waves, and to reduce the volume of water that would otherwise be thrown on to the promenade.

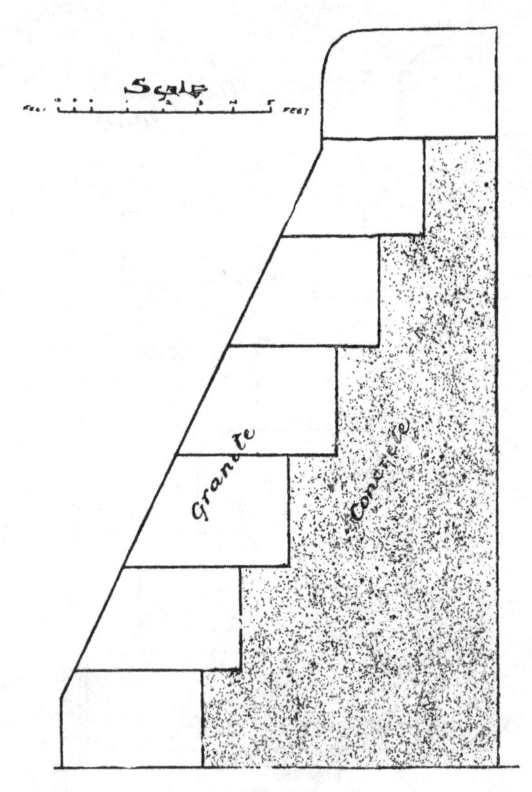

Fig. 16

Another feature in this form is, that the concussion of the waves against its face is reduced by the air that is contained in the angular spaces between each step, which forms a cushion to the force of the sea. These walls are built on a chalk foundation, and no undermining has taken place. No damage was done to these walls by the action of the heavy seas in 1897.

Mr. R. F. Grantham, in his excellent paper before referred to, said, regarding the best form of sea-wall:—"A wall built in steps, such as at Margate . . . while it does not prevent scour in heavy storms, appears to the author to be the best form. The body of water is broken as it strikes the wall, so that it is not projected upwards to the same extent, and the return to the shore is retarded, so that scour is diminished."

Fig. 16 shows the construction of a good form of sloping wall.

EMBANKED ROADS ON SEA COASTS.

Embankments are sometimes resorted to. The slopes are constructed of tenacious clay or other material that will best resist the action of the water. A mixture in a semi-wet state of coarse and fine gravel, sand, and clay or earth constitutes an excellent material when proportioned with more than half the bulk of gravel. Ordinary road scrapings and gravel mixed also make a good material for facing the slopes. The bank should be formed with a slope of 5 base to 1 perpendicular up to the extreme height affected by the sea, after which the slope may be steeper. The faces of these slopes are sometimes protected at the foot by the beach, whilst higher up, where the larger waves, during rough seas, would break on to the slope, it is desirable that they should be paved with stone. In exposed positions it may be advisable to pave the whole of the slope with stone up to the highest point affected by the sea (Fig. 17). The upper portion of the surface is usually cultivated with sea-water plants and grasses, the roots of which being of much assistance in maintaining the slope intact.

At places where the beach is plentiful the embankment slopes are considerably protected thereby, and are maintained at a nominal cost.

With regard to the protection of slopes, Mr. John Paton, M.I.C.E., in his paper on "Reclaiming Land from the Sea," read before the Inst. C.E., says:—"Where every ordinary tide reaches the foot of the dyke and there is a low exposed 'watt' or outer ground, the works comprise: (1) Pitching the sea slope with stone; (2) pitching only the foot of the dyke, and increasing the slope; (3) covering the sea face with straw matting; (4) protecting the dyke with fascine and hurdle works."

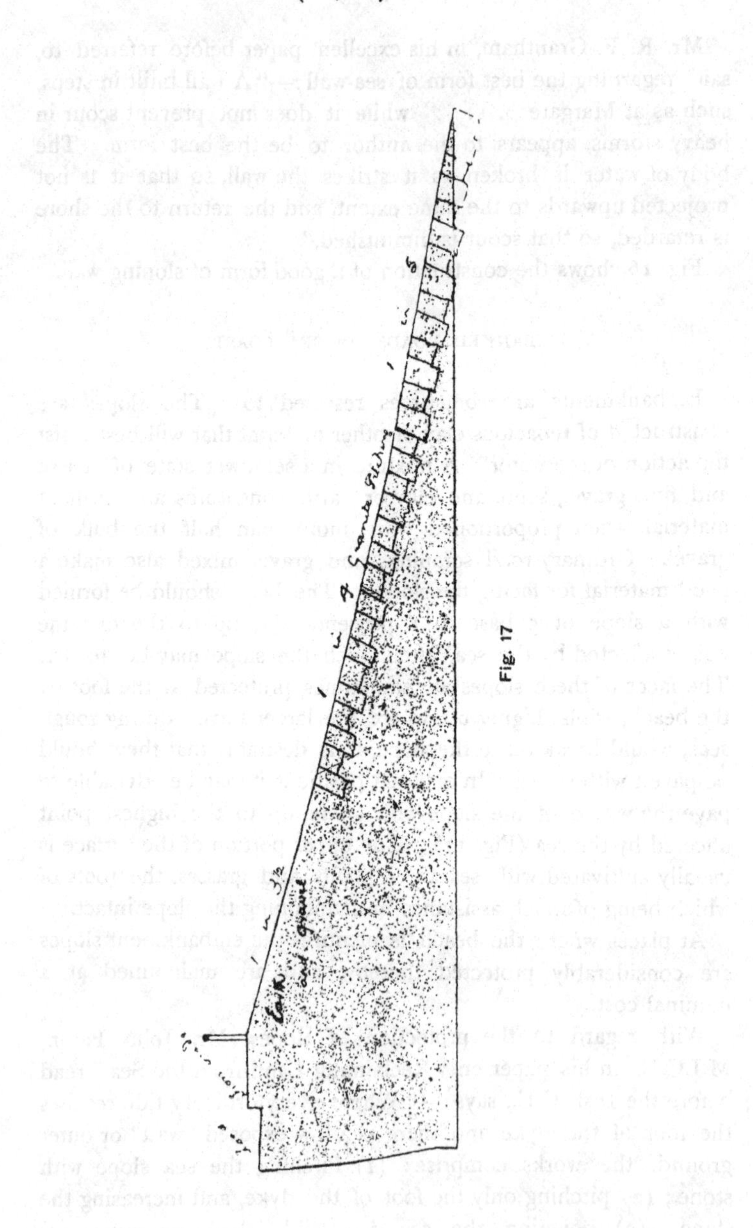

Fig. 17

His valuable remarks on these works are as follows:—

"In the first instance alluded to, of pitching with stone, the sea slope is covered with a layer of clay and small stones about 1ft. in thickness, on which are packed stone blocks of at least 300 lb. weight, the smallest section of the stones lying uppermost. When constructed on sound ground this form is almost imperishable, and affords a complete resistance to the greatest storm floods, and also receives but little damage from the ice thrown against it in the winter. Much steeper slopes are used with the solid form; generally, an inclination of 2 or 3 to 1 may be adopted. The pitching is carried to a height of from 12ft. to 16ft. above ordinary flood. Where the stones are procured with difficulty, or in cases where the ground may be equally favourable, the curved form is used, or the foot of the dyke only is covered. . . . Here the slope is lengthened, and assumes the general character of the dykes, having a considerable foreland. It answers well under certain conditions, naturally not causing so great a depression at the foot as with the stone-faced dykes.

"In the third case—protecting the slopes with twisted straw bands—it has been found that, although they have some disadvantages, particularly as regards their durability, they are considered in the end most generally advantageous. They are used first in urgent circumstances, such as in winter, or where loss of time would increase the danger or the loss of earth; secondly, when grass sods are not to be procured, and the material is clay, or sand with a considerable portion of clay; and thirdly, to repair all damages under ordinary flood. In all cases where the earth is not covered with grass the straw must be spread over as thickly as possible. The rye, or wheat straw, 2in. or 3in. thick, is fastened to the earth with straw bands, formed during the work by stitching down, perpendicular to the spread straw, the one end of a portion of the straw about 1in. thick; the loose end is then turned round and stitched down again, about the middle, into the dyke 6in. or 9in. The number of stitches is three or four per foot in both directions. When there are four per foot, the distances between the straw bands are first made two to a foot, and then care is taken that the two new stitches come between the first. Other straw ropes are then fastened and interwoven together, and the same operation is continued, the entire face of the dyke being sometimes covered in this way, the ropes being laid diagonally along the bank. As this straw matting can only be used with soft materials, it has to be renewed at least once a year, but generally twice, and very often three times. In the latter case, where the situation is exposed, the yearly expense is so considerable that the construction of a stone dyke is preferred, and is considered more economical. The average cost, including work and materials, is 3s. for 256 Danish square feet, or about 1¼d. per square yard English.

"The fascines generally used consist of brushwood bound together in bundles; the best are composed of willows. They have a length of about 10ft., a middle thickness of 1ft., and are bound. . . ."

One of the methods of using fascines, described by Mr. Paton, is as follows:—"Another method of covering the slopes, and of protecting the foot, consists in laying fascines with the butt ends at the foot of the slope, at distances of 1ft. 6in. from each other; the brushwood is then divided, giving an equal thickness on the dyke of 6in. When this is done, strong willow bands of fascines are laid obliquely at equal distances of 1ft. 6in., ascending from the foot of the slope, and are fixed by strong stakes into the dyke, the intervening spaces being filled up with stones. The slope is sometimes covered with clay and gravel."

On this same important question of the protection of embankment slopes, Mr. John H. Muller, C.E., gives the following valuable information in his paper read before the Inst. C.E., on "Reclaiming Land":—"The materials employed for the defence of slopes in Holland are of three different kinds, clay and grass flags, wood, and stone. When banks are constructed on salt marshes the body consists of clay taken from the adjoining excavations, the soil of the best quality being reserved for the outside. The body of the bank and the slopes should be well trodden down by horses, and when consolidated should be sloped afresh, and sown with a mixture of salt marsh, meadow grass, and clover seed, and the whole slope should be protected by a 'crammat.' The crammat, which costs 3d. or 4d. per square yard, is composed of a layer of clean barley straw, about 2in. thick, evenly laid, and fastened to the clay by straw bands, or strands, sixty to ninety stitches being made per superficial yard. In two or three years the bank is so consolidated that the mat does not require renewal. For banks on a lower level than the salt marsh a protection of clay and grass is insufficient. In such cases, a layer of clay, protected by stone, at a slope of 4 to 1 or 6 to 1, is employed in England, but without a cess or bench. This affords the requisite strength; but it is expensive, and as usually constructed, it needs much repair."

Mr. Muller's description of fascine work may be useful and interesting. He says:—"The wood, or fascine work protection used in Holland, consists of successive horizontal layers of fagots, 5in. or 6in. in thickness, placed in a direction up and down the slope of the bank, the thick ends overlapping the thin ends of the lower rows. Through these stakes about 4ft. in length are driven at intervals of 14in., in rows parallel to the direction of the bank. The rows are 16in. apart, and fastened down by stakes, which are left 8in. above the fagots and are connected together by means of willow binders, or 'wattles,' something like hurdle work. When this is complete, the whole is driven down with a mallet, and every fourth or fifth stake is provided with a key to prevent the wattles from lifting. If the proper sort of wood is obtained, this protection will endure from five to seven years, and is quite able to resist the action of the tide. The strength of this kind of protection might be augmented by increasing the number of stakes and binders, or by filling in with stones firmly wedged between the rows of stakes. In the application of this class of protection in England a great improvement has been effected by creosoting the stakes, but the general results have by no means been so satisfactory as those which have been obtained in Holland."

Mr. W. H. Wheeler, M.I.C.E., in his paper read before the Inst. C.E., on "Fascine Work at the Outfalls of the Fen Rivers," remarks:—"The fascines are made of thorns cut from the hedges, tied in bundles with tarred rope, the extreme length of each fascine being 6ft., and the girth 3ft. The branches, being small and tough, become interlaced. The silt brought up by the tides is rapidly deposited in and at the back of the work, and thus a solid embankment is formed of sufficient tenacity and strength to withstand the strongest tidal current, and so compact that, when necessary to remove any of the work, it can only be done by cutting the thorns out branch by branch."

PROTECTION.

One of the most important desiderata in the protection of sea walls or embankments is that the beach or sand should be retarded or prevented from being washed or removed from the shore in the vicinity of the defence works; every effort should be made to add to the accumulation, so as to form a natural bar, to check the force of the sea, and for the protection of the foundation of the sea wall. In order that these accumulations should be collected and retained, some artificial means, such as groyning, are generally necessary, and the providing of such works, when practically applied, effects considerable economy in the long run.

Various descriptions of groynes are used, and their success is more or less dependent on the set of the tide, more particularly in rough weather. Most coasts are attacked by the sea to a greater extent when the wind is in one direction than in another, so that every coast has its more or less destructive wind. These winds may drive the most scouring seas against promenade walls at different angles, according to their relative aspect to the prevailing gales. The form of groyne adopted should for these reasons be planned to give the best results according to the particular conditions surrounding each part of the coast.

Many forms of groynes have been designed, but the most useful and successful of them has been the invention of the late Mr. Edward Case, C.E.

The author has been much interested in Mr. Case's invention from the beginning, when he put down his first experiments to protect the old Dymchurch wall.

Although Mr. Case received little encouragement at the hands of his critics, he persevered in his work, and at length succeeded in accumulating a fine depth of sand, entirely protecting the foundation of the sea defences under his care. Mr. Case's system, which he afterwards patented, differs from most of the forms, inasmuch that, instead of constructing groynes across the beach, with a gradual slope to reach high-water mark, he commenced with low ones like a sill, beginning at mean sea level and extending down to low-water spring tides. Mr. Case's experience convinced him that the most extensive travel of beach and sand was between mean sea level and low-water mark. His object was to build up the shore from extreme low-water mark, so as to give as flat a gradient to the beach as possible, and so obtain a natural angle of repose to the accumulating material.

Some valuable remarks on groyning are given in Mr. Case's paper on "The Dymchurch Wall," which was read before the British Association in 1899. It is claimed that Mr. Case and his firm (Messrs. Case and Gray, of Westminster) have invariably obtained successful results wherever they have been consulted, and their schemes fully carried out, both on the various shores of the United Kingdom and abroad.

Mr. R. G. Allanson-Winn, C.E., who has done much to make known the value of the correct application of this system of groyning, says, in a paper read before the Institution of Civil Engineers, Ireland:—

"Mr. Case took a long step in advance when he promulgated his theory that every shore had its own inclined surface of repose, and that that surface could be brought about by accumulations mainly derived from the travelling material between mean sea level and low-water mark. 'If,' he argued, 'a shore can be made to at once acquire this surface (*from which there is a minimum chance of alteration through the action of waves and currents*), sea walls, properly so-called, will not be so necessary as formerly for the protection of esplanades, sea-side roads, &c. If a good 'full' of beach can be secured, a comparatively light retaining wall will be sufficient, and we shall be able to rely on the beach itself as a protection for our foundations, whilst we shall have the satisfaction of having saved heavy expense in costly concrete blocks, &c., which would otherwise have been considered necessary."

Mr. Case's patent groynes can be set in any direction or in almost any form; they are inexpensive, easily constructed, and are far less obstructive to pedestrian traffic on the beach, and more agreeable to the sight than most other forms of groynes.

These groynes are so made that they may be very easily heightened as the beach collects.

It is evidently an advantage to construct groynes comparatively low at first, and to add to their height as the beach accumulates; the groyne should, further, be continued fully down to low-water spring tides, and the planking carried to the depth of any shifting sand or beach. The results produced by the adoption of high groynes are often a source of danger rather than protection to a sea wall. Instances can be frequently seen around the coast, where such groynes are carried out, of deep scours taking place on the lee sides during storms, which gaps are only too sure to create a terrible force at perhaps the weakest part of the wall at times of heavy seas from an opposite direction.

Mr. Robert Pickwell, in his paper on "The Encroachment of the Sea," read before the Inst. C.E., refers to some groynes erected by himself at Withernsea in 1870-71, which were attended with great success. He says: ". . . . To oppose as little resistance to the breakers as possible, the top five rows of planks were only added as the beach accumulated." He also remarks with reference to the groynes carried out by Mr. W. M. Jackson at Hornsea: "The sheer piles being fixed at their ultimate height (about 9ft. or 10ft.) above the surface of the beach, present too great a resisting surface to the breakers, and are more exposed than that form of groyne which admits of the top planks being added and 'built up' as the beach accumulates."

CHAPTER VI.

DRAINAGE.

THE effectual drainage of a road is a question of much importance, requiring a good deal of attention, as no road can be constructed to give satisfaction unless provision is made for its drainage. In order to preserve a road when once made, means must be provided for carrying off the surface water and for maintaining the foundation as dry as possible. Nothing is more destructive to a road than wet, especially during frosty weather; with this in view, therefore, the best means should be taken to make a dry foundation and a dry surface. The foundation or bed, which generally consists of earth, chalk, or rubble, must receive a good deal of attention, as it is the basis of the work. The carrying strength of the road surface is principally governed by the condition of the foundation upon which it is constructed. It is therefore a matter of first consideration, in order that the road surface may be firm and even to carry the loads that pass over it, that the road bed should be firm and dry, and free from subsiding and incoherent substances.

The means commonly adopted for sub-drainage are by constructing drains to convey the water away, and sometimes where the site, owing to the levels of adjoining lands, is below any means of outlet for such drains, then by raising the road as much as may be necessary in order to be above the level of the water. Where drains are resorted to, these should be constructed so as to carry off as much as possible of the land water before it reaches the road. These drains can either be open roadside trenches or covered soaking drains (Fig. 18). In situations where proper sewers exist this sub-drainage may more easily be dealt with by means of side drains carried under the water channel. It is, however, not always necessary to construct sub-water drains for roads; it is only in places where the ground is usually damp from want of natural drainage or from other circumstances which have a tendency to create wetness. Covered sub-drains (Fig. 18, B and C) are to be recommended in preference to open ditches, owing to the latter being liable to overturn vehicles, and to the trouble experienced in the sides of the trenches falling in and being washed away.

A properly-constructed covered drain should admit of water being readily received into it, without becoming clogged with the gravel washing into the openings provided.

Tile drains (Fig. 18, B) are best adapted to meet these requirements. These consist of a trench, at the bottom of which is laid

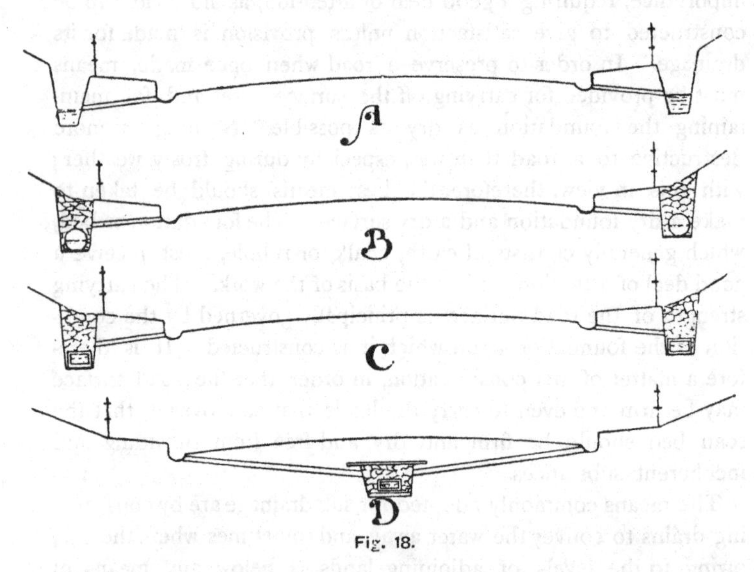

Fig. 18.

a series of open-jointed agricultural drain pipes or tiles. Upon this being done the trench is filled with rounded stones 6in. in diameter at the bottom, becoming gradually smaller toward the top. In constructing a drain of this description care must be exercised to properly compact the pipe underneath and at the sides with suitable loose stones, as otherwise the superincumbent weight would be likely to crush the pipes. The box drain (Fig. 18, C and D) is one generally adopted in districts where stone is plentiful, as it is constructed entirely of stone. This form of drain only differs from that of the one already described in the detail that, instead of using unjointed pipes, flat stones placed together, so as to form a rectangular drain, are laid at the bottom of the trench.

In cases where much water is to be conveyed away by means of a porous drain, especially on inclines, it is desirable to protect the bottom of the trench by a layer of concrete, otherwise the foundation will be washed away and the surface will subside.

Besides the necessity of sub-water drainage, there is much necessity for the water falling on the surface to be immediately conveyed away. It is most important that no water should be allowed to remain upon the surface of a road, the principal danger being that of its tendency to percolate into the road, this being liable to cause disintegration and subsidence. The carrying off of the water is generally provided for by open channels carried down each side of a road. The means adopted of turning the water off the surface into these side channels is by barrelling the road in cross section.

The section for this purpose is sometimes formed with a flat camber, and at others with two practically flat surfaces brought together with a rounded centre to form the crown. Care must be taken not to make the slope too great, or the road may be dangerous to horse and vehicular traffic, especially in cases where smooth materials are used on the surface. The camber allowed by different engineers varies, but for a macadamised road 1 in 30 to 1 in 40, according to the hardness of the stone used, and 1 in 45 for wood pavement, and 1 in 55 for asphalt, are good averages. Mr. Macadam said, in regard to this question: "I consider a road should be as flat as possible with regard to allowing the water to run off at all, because a carriage ought to stand upright in travelling as much as possible. I have generally made roads 3in. higher in the centre than I have at the sides when they are 18ft. wide; if the road be smooth and well made, the water will run off very easily in such a slope."

Mr. Telford says: "The surface (of the road itself) should be made with a very gentle curve in its cross section, just sufficient to permit the water to pass from the centre towards the sides of the road; the declivity may increase towards the sides, and the general section form a very flat ellipses, so that the side, at the time, should, upon a road of about 30ft. in width, be 9in. below the surface at middle." Mr. Walker says (Parliamentary Report, 1819): "I consider a fall of about 1½in. in 10ft. to be a minimum in this case, if it is attainable without a great deal of extra expense."

The water thus turned off the road's surface is conveyed by means of side channels for some distance, and, at intervals, outlets are constructed in the line of the gutter. Outlet drains to convey the surface water from the side channel should be placed at every point where water is inclined to lie, owing to the fall of the road; similar openings should also be provided at frequent interval at the sides of hills, the distance apart to be judged according to the gradient; but the maximum distance should be 75 yards, so as to guard against the road quarters beyond the channel being cut by the overflowing water. At the inlet of each

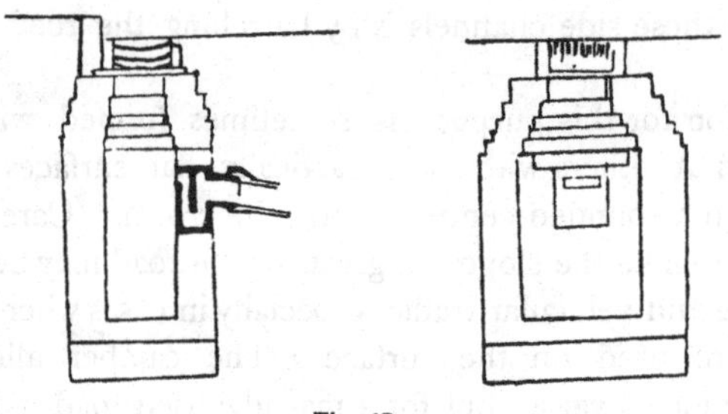

Fig. 19.

of these drains a catch-pit should be constructed with grating over, to prevent as much as possible of the gritty sediment entering them.

In towns it is also necessary to trap these openings by means of a proper trap or syphon, to prevent disagreeable odours escaping from the drains and sewers. Fig. 19 shows a catch-pit constructed of brickwork with a Doulton's trapped gully-block built in one side, from which the drain connection is made.

Fig. 20 illustrates Sykes' Patent Street Gully, manufactured by the Albion Clay Company, Limited. The advantages claimed for this gully are:—The water descends and rises again over a diaphragm before it escapes, which action arrests all road detritus. Paper, straw, &c., rise to the surface and cannot enter the sewer. The deep seal prevents the untrapping in hot weather. The

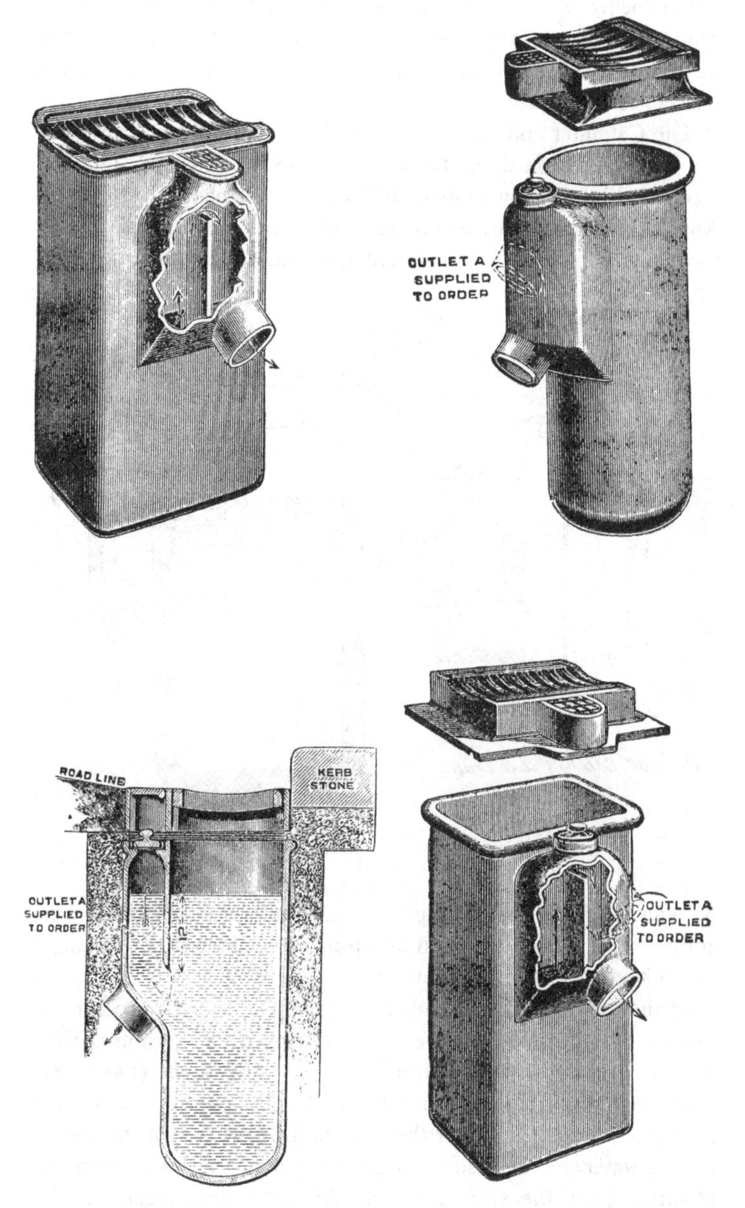
OUTLET A
SUPPLIED
TO ORDER
ROAD LINE
KERB
STONE
OUTLET A
SUPPLIED
TO ORDER
OUTLET A
SUPPLIED
TO ORDER

Fig. 20.

outlet being well below the road cannot be broken by steam rollers. It affords easy access for cleansing through screw-plugged inspection eye, and being made in strong stoneware or iron it resists great pressure.

The Catch-pit and Cascade Trap (Fig. 21) is manufactured by the Patent Victoria Stone Company. The whole is made in one piece, and is strongly constructed of the patent Victoria stone. Any sticks, straw, or other floating substances that may pass into the gully are prevented from entering the sewer and can easily

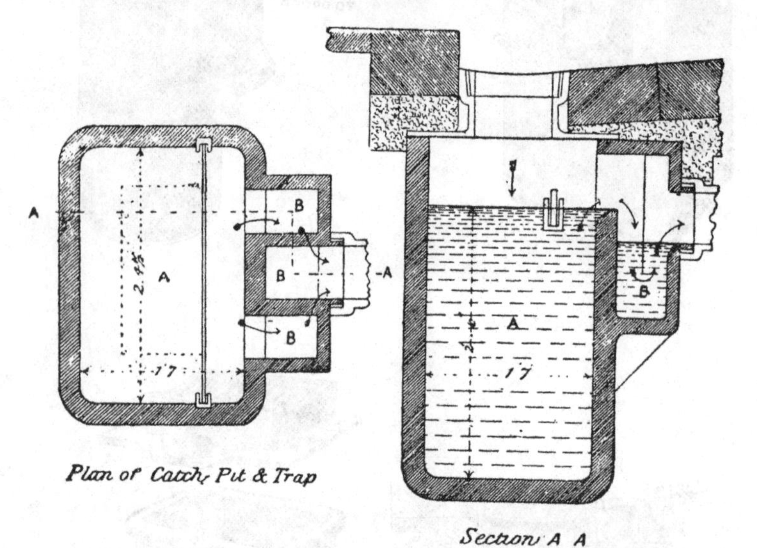

Fig. 21.

be removed from the catch-pit, whilst another good feature in the gully is that there is a good fall from the catch-pit to the trap, which keeps the latter well flushed.

In the case of country roads it is, as a rule, sufficient to simply carry a drain from a square pit constructed of masonry, with grating over the water channel, into a side ditch (Fig. 18). Where a road has little or no fall in its length, endeavours should be made to raise it at either the ends or the middle, or by lowering the water channels, so as to give an inclination, for the purpose of carrying off the surface water. All particulars regarding the construction of sewers and drains in streets are to be found in Mr. Baldwin Latham's valuable work, "Sewerage."

When dealing with the question of channelling, it is generally necessary to include that of kerbing; as a rule channels are incomplete without proper kerbs. The kerb not only forms a sill against which the paving material of the footpath abuts, but it constitutes the side of the water channel by which the water is prevented from running over the footpath (Fig. 22).

According to the Model Bye-laws of the Local Government Board such kerbs "shall be not less than 3in. at the highest part of such channel, and not more than 7in. at the lowest part of such

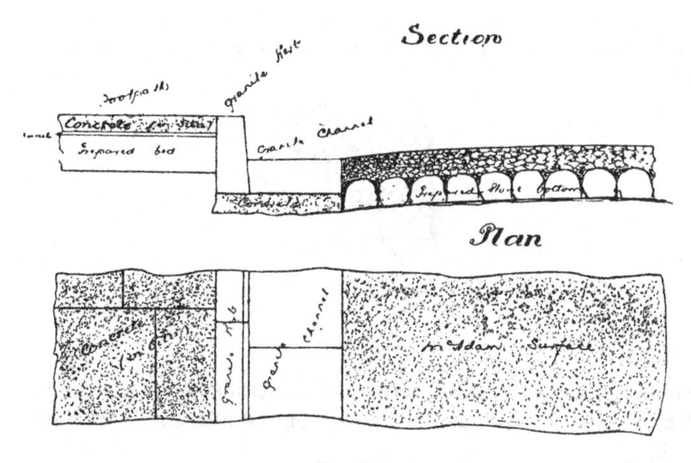

Fig. 22.

channel." Kerbs constructed too low have disadvantages, and also when too high. In the former case it would in many places admit of the surface water overflowing the footpath, and there would also be an inclination for vehicles to be driven on to the footway, to the danger of pedestrians; whilst in the latter, if the height exceeded 7in, it would be inconvenient, and perhaps dangerous to pedestrians.

The materials most suitable for kerbing and channelling are granites and syenites. These materials, owing to their durability, are very extensively employed; in fact, their use in streets with heavy vehicular traffic is almost indispensable, as no other material would stand for any length of time the rough usage to which they are subjected.

Other materials, such as sandstone, Kent rag, cement-concrete, either laid in blocks or *in situ*, vitrified fire-clay blocks, &c., are frequently employed for either kerb or channel, or both. These materials cost much less than the granite, &c., but are not to be recommended in streets of much traffic; in fact, tneir use often results in as much money being spent in repairs and renewals as would in the first place have paid for the superior material.

Cast and wrought iron kerbs have been used both in this country and abroad, and when properly made they have a very good appearance (Fig. 23). When cast with a roughened foothold for pedestrians these kerbs are adaptable to all classes of streets, but unless this feature is effective they wear very slippery, and are a

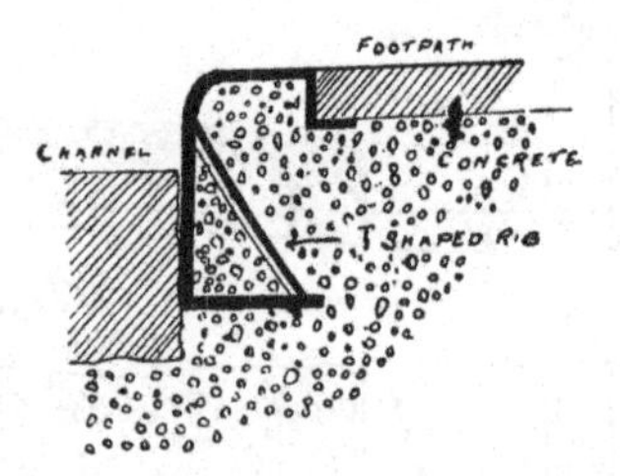

Fig. 23.

source of danger. These kerbs are sometimes cast in a form which combines a kerb and channel in one, similar to a trough set on side.

Kerb stones are usually of the following sizes:—12in. by 8in. laid flat, with surface splayed to the slope of the footpath; 6in. by 12in. laid cn edge; 9in. by 4in. laid on edge. These stones should not be in less lengths than 3ft.

The top surface of the kerb should be hammer-dressed, and the front face should be similarly treated, in order to give a smooth surface against the gutter and channel stone, and the back should be also dressed for 3in. from the top, so that the paving material will be laid fair against it. Some kerbs are drafted with the chisel for a depth of an inch or so along the edges, so as to give a better finish and appearance to the arris of the stone. At joints the stone should be chiselled 6in. deep at right angles with the top and side faces of the kerb, and all projections beyond this depth should be removed, so that a close joint may be made. The face of kerbs should further be slightly bevelled, so that when vehicles are drawn against them the grinding action will be less, and the damage to the paint on carriage wheels will be minimised. In setting the kerb, much care is necessary to secure a sound and solid foundation. Unless this care is exercised the kerb is sure to either sink or fall forward, especially at times when the road is being repaired and rolled by a heavy roller. In order to obviate this trouble, the safest plan is to set the kerb on a bed of good concrete.

Stone, either in slabs or blocks, is usually employed for the gutters in macadamised streets. The water table should not be less than 18in. in width, and should be laid with the same slope as the roadway. The stone should be properly dressed, as in the case of the kerb. The length of channel slabs should not be less than 3ft, and their thickness 4in. to 6in., or even more for heavy traffic. Channel blocks and slabs should in each case be set on a bed of good concrete, or the process of rolling, or the effect of heavy traffic, will soon cause settlement, resulting in the formation of puddles of water and mud. Where a street is paved with wood blocks or granite setts, &c., the channel is paved with the same description of material, with the blocks laid with their lengths parallel to the kerb.

CHAPTER VII.

FORMATION OF ROADS.

BEFORE considering the system of formation best suited to carry the road surface, it is as well that some reference should be made to the methods adopted by the old successful masters. Mr. Telford's system for broken stone roads was to construct a foundation to carry the road's surface, by means of a regular pavement of stones (Fig. 24). The following is a specification quoted from Sir H. Parnell's work for a Telford road foundation :—"Upon the level bed prepared for the road materials a bottom course or layer of stones is to be set by hand in form of a close, firm pavement;

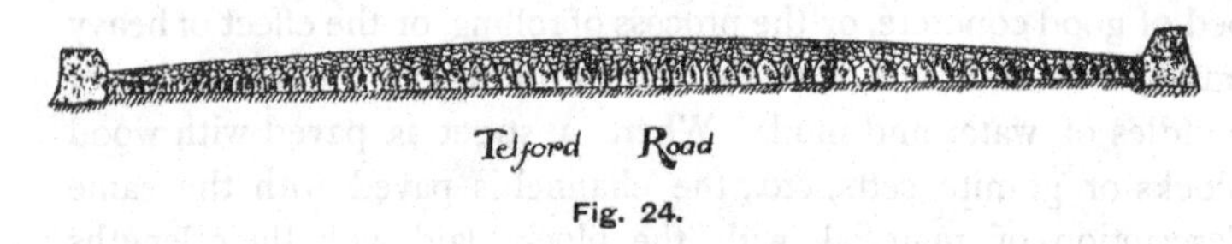

Telford Road

Fig. 24.

the stones set in the middle of the road are to be 7in. in depth, at 9ft. from the centre 5in., and 12ft. from the centre 4in, and at 15ft., 3in. They are to be set on their broadest edges lengthwise across the road, and the breadth of the upper edge is not to exceed 4in. in any case. All the irregularities of the upper part of the said pavement are to be broken off by the hammer, and all the interstices to be filled with stone chips firmly wedged or packed by hand with a light hammer, so that when the whole pavement is finished there shall be a convexity of 4in. in the breadth of 15ft. from the centre." These foundations would have the appearance of a rough pitched pavement, with stones set on end, with interstices filled with smaller stones. This pavement thus formed a very substantial and firm foundation for the top surface to be placed upon.

Mr. Telford's method differed but little from that which had been adopted in France many years before by M. Tresagnet. M.

Tresagnet's road is shown in Fig. 25, in which it will be observed that, whilst the stones in Telford's foundation rest on their flattest ends, and their sizes are graduated from the centre to the sides, in the former the stones are placed on edge on a bed formed to the same camber as that required for the finished surface. Fig. 26

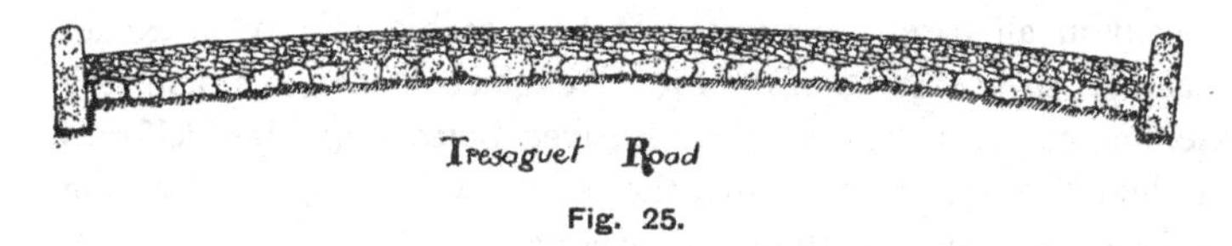

Fig. 25.

shows the method adopted in road foundations in France previous to Tresagnet's system 1764.

Mr. Macadam's system differed from the former methods in respect to the foundation. He contended that as long as the road bed was kept in a dry state, by proper attention to drainage, &c.,

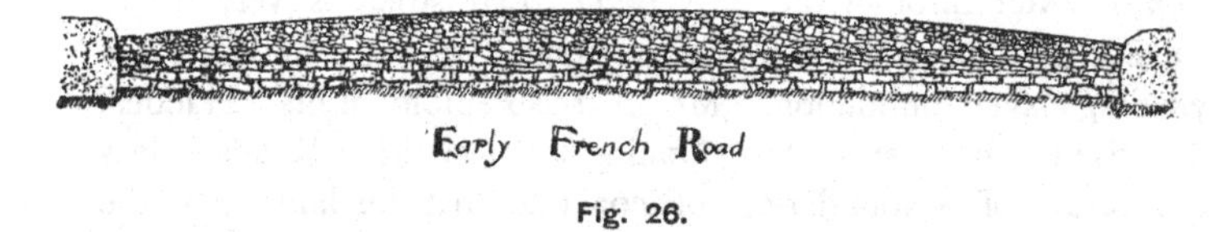

Fig. 26.

it would carry the traffic with no danger of giving way. The road surface therefore had to be constructed impervious to water, and Mr. Macadam considered that the thickness necessary for the road covering was that which would effect this requirement. He further contended that a hard foundation had a tendency to cause

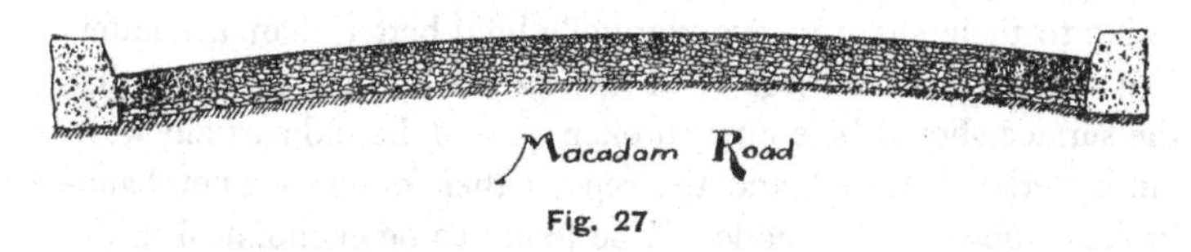

Fig. 27

the road to wear away sooner than when the surface was placed on a soft substance (Fig. 27).

However sound Mr. Macadam's doctrines may appear to be from his evidence before the Committee of the House of Commons, engineers are mostly convinced that the last contention requires a good deal of modification, inasmuch as it is, as Sir Henry Parnell says, "contrary to the first principles of science."

It is apparent that the question as to whether or not the natural bed is sufficient in itself to carry the road with its traffic is one which must be governed by local conditions, which will vary at different places, so that no rigid rule can be laid down to apply equally in all cases. Professor Mahan, in his "Essay on Road-making," says: "Some of the French engineers recommend, in very yielding clayey soils, that either a paved bottoming after Telford's method be resorted to, or that the soil be well compressed at the surface before placing the road covering."

There can be no doubt but that solid chalk forms an excellent road foundation for the reception of Macadam's system; but in those districts where clay is the principal formation this method may fail to give satisfaction. It is almost a practical impossibility to maintain a broken stone road covering impervious to water, the consequence being that when the bed is softened by the penetration of water through the surface the road stone is very apt to subside by being pressed under the weight of the traffic into the yielding clayey foundation. Mr. Charles Penfold, in his "Practical Treatise on the Best Mode of Making and Repairing Roads," gives the result of a foundation of concrete that he laid upon the Walworth-road. He says: "It was raised by 9in. of concrete and 6in. of granite and Kentish rag stone mixed, and in some parts it was covered by rag and flints The improvement is so great with respect to the annual repair, that the Trust directed it to be applied to upwards of two miles of road upon which the greatest traffic exists" Upon weak foundations it is better to use hand-broken stones in preference to machine-broken, as the former, owing to their sharp angles, naturally bind better than the latter. It is, further, an advantage when such a road, is being repaired, that the surface should be entirely broken up and the old partially worn out materials removed, and the repairs then done with new hand-broken stones. This mode will be found to be economical in the long run. The materials removed may be sifted and freed from all useless stuff, and used again in streets requiring less attention.

The usual method of breaking a road by hand labour, and scoring at intervals of every 5in. to 8in., is unsatisfactory in practice unless the foundation is perfectly to be relied upon, for otherwise the new covering will settle at the cross cuts, through not being so firmly supported as it is at the ridges. Inferior foundations are more often than not to be fonnd in connection with embanked roads. These foundations should be made solid, even though the cost may be considerable; otherwise is false economy. Weak foundations mean endless repairs, the disconcerting of the traffic, and inconvenience to the public.

In clayey foundations the soil removed in forming the road may be used as a foundation material after proper preparation. The material is usually piled up into heaps, and burned into ballast, and spread over the road bed as bottoming.

To carry a foundation over very soft soils bundles of long, thin sticks or faggots are sometimes employed. These will form a good foundation if placed in such situation, or at such a depth as will cause the sticks to always remain wet; otherwise they will quickly become rotten, and perhaps form a more defective bottom for the road than the one they were intended to improve. A good and dry foundation for clayey sites is constructed by means of stones, about 4in. cube, with their inter stices filled in with cinders. It is, however, a mistake to use too many cinders in this work, as they do not readily bind together. Old tins well beaten flat and placed on embanked roads, and steam rolled, make an excellent foundation; but the drawback to their use is that should it be necessary at any time to open the road to lay new pipes, &c., an unusual area of surface is generally disturbed and damaged, as the layers of tins lap one over the other, and are difficult to part.

CHAPTER VIII.

ROAD COVERING.

Having secured a good road bed and foundation, the next work is to construct a suitable covering that will be as impervious as possible to water and produce a suitable and durable surface to carry the traffic. This subject may be divided up into five sections, viz.:—Broken stone roads, and pavements constructed of wood, asphalt, stone, and brick.

The first of these, viz., broken stone roads, are the most generally adopted. The material adopted should be as tough as possible and of uniform hardness. The fact of a stone being hard and of great resisting strength to crushing is not always a sign by which to judge its wearing qualities. Some stones are hard and brittle, and pound away to dust in a short time; while others may be less hard, yet wear for a greater length of time; whereas others, again, may be too soft and crush, and prove very wasteful. The stone used must, further, be such that will least be affected by wet, frost, and atmospheric changes.

Regarding the relative strength and durability of various road materials, Mr. Thomas Codrington says: "It is a difficult matter to determine. No test but actual wear in the road can be fully relied on, and though it is easy to see that one stone wears twice or three times as long as another, it is almost impossible to take into account all the circumstances under which they are exposed to wear. The nature of the traffic has a considerable effect on the relative wear, as well as on the actual wear of different materials, and the moisture or dryness of the road has often a great effect on the wear of the same material."

The engineers of the French Ponts et Chaussées have endeavoured to arrive at a comparative numerical value of the qualities of the materials used on the national roads, and "co-

efficients of quality" are given for the various materials used in each department.

The following list has been compiled from a return for 1876 :—

Coefficients of Quality of Road Materials.

Granitic gravel	23·8
Quartz gravel	21·4
Trap	20
Quartz	10 to 25 (in one instance 4·8)
Basalt	12 to 20
Porphyry	10 to 20 (in one instance 5)
Quartzite	11 to 18
Devonian schist	16
Schist	4 to 12
Sandstone	12 to 16
Granite	6 to 20 (generally 10 to 12)
Syenite	12
Gneiss	9 to 12
Silicious pebbles and gravel	8 to 19 (in one instance 6)
Silex	8 to 16
Chalk flints...	7 to 11·6
Silicious limestone	6 to 18 (generally about 10 to 12)
Compact limestone	14
Magnesian limestone ...	12
Carboniferous limestone ...	9
Oolitic limestone	5 to 12
Lias limestone	5 to 10
Jurassic limestone	5 to 8
Limestone	5 to 12
Mean of all France	10·63

The life of road stone may be increased by seasoning; stone, like wood, when first removed from its natural formation, is green and unseasoned; it is therefore desirable that a stock of stone should be quarried and broken the year previous to use, so that it may be exposed to the air for some time before being laid on the roads.

The following is quoted from the "Encyclopædia Britannica":—
"Roads and Streets.—Coefficients of quality for various road materials have been obtained by the engineers of the French Administration des Ponts et Chausees. The quality was assumed to be in inverse proportion to the quantity consumed on a length of road with the same traffic, and measurements of traffic and wear were systematically made to arrive at correct results.

These processes requiring great care and considerable time, direct experiments on resistance to crushing and to rubbing and collision have also been made on 673 samples of road materials of all kinds. The coefficients obtained by these experiments, which were found to agree fairly well with those arrived at by actual wear in the roads, are summarised in the following table. The coefficient 20 is equivalent to 'excellent,' 10 to 'sufficiently good,' and 5 to bad.'"

Material.	Coefficient of wear.	Coefficient of crushing.
Basalt	12·5 to 24·2	12·1 to 16·00
Porphyry	14·1 to 22·9	8·3 to 16·3
Gneiss	10·3 to 19	13·4 to 14·8
Granite	7·3 to 18	7·7 to 15·8
Syenite	11·6 to 12·7	12·4 to 13·00
Slag	14·5 to 15·3	7·2 to 11.1
Quartzite	13·8 to 30	12·2 to 21·6
Quartz ore sandstone	14·3 to 26·2	9·9 to 16·6
Quartz	12·9 to 17·8	12·3 to 13·2
Silex	9·8 to 21·3	14·2 to 17·6
Chalk flints	3·5 to 16·8	17·8 to 25·5
Limestone	6·6 to 15·7	6·5 to 13·5

SIZE OF ROAD STONE.

The stone for macadamising a roadway should be properly broken to a gauge that will give the best results, according to the material employed.

Hard stone should be broken rather smaller than weaker descriptions, as hard stone is better able to resist the weight of the traffic than the latter.

The maximum size to which any stone suitable for road-making, and intended to be used on the surface, should be broken, is one that will pass in every direction through a ring 2in. in diameter. It will be found an advantage to break very hard stone to a smaller gauge than 2in. Mr. Macadam, in his report of 1810, says, regarding the size to which road stone should be broken :—"The size of stones for a road has been described in contracts in several different ways, sometimes as the size of a hen's egg, sometimes as half a pound weight. These descriptions are very vague, the first being an indefinite size, and the latter depending on the density of the stone used, and *neither* being attended to in the execution.

The size of stone used on a road must be in due proportion to the space occupied by a wheel of ordinary dimensions on a smooth, level surface; this point of contact will be found to be longitudinally about an inch, and every piece of stone put into a road which exceeds an inch in any of its dimensions is mischievous."

Mr. Thomas Codrington, in his valuable work, "The Maintenance of Macadamised Roads," says:—" . . . but a stone that would pass in its largest dimensions through a 2½in. ring, soon became and has since remained a recognised size. A cube of rather less than 1½in., containing three cubic inches, and about 6 oz. in weight, will pass this gauge, and it is a size sufficiently small for road-making. For surface repairs a smaller size may be used with advantage, especially when the material is hard, as it covers a larger surface, consolidates sooner, and makes a smoother road. The metalling called 'medium' in London, not exceeding 5 oz. in weight, will pass through a 2in. ring, and a 4 oz. size is sometimes used. . . . The tougher the stone is the smaller it may be broken, with advantage to the road, but, of course, at an increase of cost. On roads where the traffic is of a heavy character, breaking the materials to a small size is an expense not attended by any advantage, and if the stone is not very tough, too large a proportion of small stuff is produced, which is of little use in the road, and can only be separated by screening, at an extra expense. Metalling to be laid with a roller need not be broken so small as it must otherwise be. When stones are once well bound together, as they are by a roller, they are the stronger for being larger."

CHAPTER IX.

BREAKING.

HAVING decided upon the size of the material to be used, the next question is, Which is the best method to adopt for breaking the stone? There are two methods by which stone may be broken for road purposes, one being by the employment of hand labour and the use of a hammer, and the other by a machine stone crusher. The first method is operated by men and boys, who, by the aid of a large iron cleaving hammer, break the very large stones as supplied from the quarry into pieces or 6 lb. of 7 lb. weight; these should then be sorted over, and all the good coloured material of good wearing appearance should be placed in one heap, and the weaker class placed in another heap. The breakers then, by means of a small napping hammer, weighing 1 lb., made of iron faced with spherical steel ends, which is securely fixed at the end of an ash handle 18in. in length, break these stones thus prepared into smaller sizes. This work being completed, the stones are employed for use according to their class, the selected stone for the most used streets, and the weaker for less important roads.

It requires a good deal of practice to become an expeditious stone breaker; the dexterous hand will break a large quantity of hard and tough stone with comparatively little exertion, the method being to strike the stone with sharp strokes, and in direction of the grain. The material is thus broken to the sizes required, and with little or no waste. The cost of breaking by hand varies, according to the hardness and toughness of the material, from 1s. to 4s. per yard cube. The other method employed is the machine stone crusher. This is a rapid process, by which the larger stone, in passing between a contrivance of steel jaws set in motion by machinery, is crushed into fragments, which in their turn pass into a revolving riddle, which is an iron cylinder perforated with holes to the size required for the road stone.

The material which has been sufficiently broken passes through these perforations, and is delivered on to a platform ready for use, while the remainder is automatically returned by means of an elevator into the crushing chamber, to be again broken.

Some authorities are of opinion that there is a disadvantage in the employment of a crusher as compared with hand breaking, in that the stone splinters and flakes more or less during the process, and the articulation of the stone in binding is less satisfactory; and, moreover, the crushing strain has a tendency to affect the texture of the material, rendering it less tough.

These unfavourable conditions do not obtain to such a large degree when the stone is hand broken.

With regard to the question as to whether the quality of the stone suffers by the crushing strain it receives in the process of machine breaking, the author investigated the matter in order to ascertain what is the truth of the matter by testing the powers of absorption of stone from three different quarries, hand and machine broken. The following are the results after a week's immersion in water:—

Polmennor Quarry, Penzance. Blue Elvan.

Hand broken—

Dry	11 lb. 20 dr.
After immersion...	11 lb. 26¾ dr.

Machine broken—

Dry	11 lb. 20 dr.
After immersion...	11 lb. 28 dr.

Kinedjackcliffe Quarry, St. Just. Basalt, hard and dense.

Hand broken—

Dry	10 lb. 2½ dr.
After immersion	10 lb. 4½ dr.

Machine broken—

Dry	10 lb. 5 dr
After immersion	10 lb. 7 dr.

Levant Quarry, St. Just. Basalt, hard and dense, but blasted for mining purposes.

Hand broken—

Dry	10 lb. 10 dr.
After immersion	10 lb. 14¾ dr.

Machine broken—

Dry	10 lb. 0 dr.
After immersion...	10 lb. 5¼ dr.

It appears from these tests that little or no injury resulted to the strength of the hard and dense stones from Kinedjackcliffe, whereas, in the Elvan, owing to its being a stone with more grain, a slight degree of disintegration had taken place, and which was also the case with Levant, which had been weakened in blasting operations.

The following are the results of practical tests made by the author with lengths of road coverings laid on equal conditions from the three foregoing quarries. In this test the stone from each quarry was laid in two lengths, one hand broken and the other machine broken, and after a year and three months' wear the road was cleansed and carefully examined by the author, accompanied by other gentlemen. Following results:—

Stone.	Surface results after 15 months' wear.	Average thickness in favour of one or the other worn in centre of road.
Elvan stone, medium strength.	Hand broken, slightly better condition on surface than machine broken.	Hand broken, worn $\frac{3}{4}$in.; machine broken, worn $\frac{7}{8}$in.
Basalt, very hard, properly blasted for road purposes.	Slightly in favour of machine broken.	Wear equal in each case.
Basalt, blasted in mining operations.	Difficult to judge the better of the two samples. Numerous pitholes in each.	A good 25 per cent. less crushed fragments to pass a $\frac{1}{2}$in. ring in the hand-broken sample as compared with that of the machine broken.

Mr. Thomas Codrington says regarding machine broken stone:—"Stone broken by machine is not so durable as if hand broken. There is always, even with the hardest stone, a certain amount of crushing which is greater when the jaws become worn. A stone not so hard, such as mountain limestone, suffers so much from the crushing as to stand very little wear on a road afterwards. The stones are not so cubical in form or uniform in size as if well broken by hand."

From inquiries made the author found that of thirty-one municipal engineers experienced in road-making, and who had used both hand and machine broken stones, twenty-five preferred the former, owing to its greater durability.

The process of breaking stones by machinery is, as a rule, less costly than the employment of hand labour, but the extent to which there is a saving in this respect much depends on the hardness and toughness of the material dealt with, and the cost of fuel. A very hard stone can be broken by a machine fitted with Hadfield's manganese steel jaws with almost as much ease as it will a softer material. But the relative cost of breaking by hand the harder material, as compared with the softer, would be twice as much.

The following were observations made by the author on two years' working, during which period nearly 4000 tons of stone were broken by machine and a quantity by hand. The cost of quarrying and breaking Elvan stone by a Baxter machine for two years' supply worked out at 4s. 3d. per ton. This figure included all charges and costs, besides 3½ per cent. per annum on capital outlay on the machines employed, spread over a period of fourteen years. There was no marked difference in the cost of breaking Elvan and that of a very hard basalt.

The cost of hand broken Elvan, including cost of raising, breaking, and all charges, amounts to 5s. per ton, or an increase of 9d. per ton over that of machine broken; the cost of breaking by hand the harder stone, taking other items of cost to be the same, would bring the cost to 5s. 10d. per ton, or about 1s. 7d. per ton greater than the cost of breaking similar stone by machine.

Mr. William Ballantine, District Surveyor, Falkirk, in 1886 made some useful trials to arrive at the cost of breaking by a Baxter's stone crusher, which resulted as follows:—The quantity broken per day of 9½ hours is from 65 to 85 tons; as much as 18 tons have been broken in one hour, but taking 65 tons as an ordinary day's work, the cost of breaking is as follows:—

Labour, 9 men at 3s. 6d. per day	£1	11	6			
Engine-man at 5s. per day...	0	5	0			
Feeders { one man at 4s. per day	0	4	0			
Feeders { one boy at 2s. 6d. per day	0	2	6			
				£2	3	0
Coal, 5 cwt. at 8s. per ton				0	2	0
Oil and tallow				0	1	0
Allow for depreciation and repairs (working six months)				0	4	0
Total				£2	10	0

which sum is about 10d. per ton, or allowing for time lost in removing from one place to another, the actual cost is 1s. per ton, as compared with 2s. 3d. per ton for breaking by hand.

In the first item nine men are employed in the foregoing calculation, five of which are for taking the broken material from the machine. In the more modern machine, however, only one man is required for this work. The cost of coal, it will be observed—8s. per ton—is very small compared with that paid in most districts.

At Alford the following year a trial was made with a new Baxter machine with automatic delivery appliance, with the following results. *Aberdeen Journal*:—" The machine was continued at work for seven hours, and during that time broke, screened, and loaded 40 cubic yards of metal at a cost of 9d. per cubic yard The price of breaking the quantity by hand is 2s. 2d. per cubic yard, besides 2d. per load for loading, being a save of 1s. 7d. per yard. The author has, it will be seen, included in his calculations the actual cost of the repairs done to the machine during the two years of breaking, together with a suitable percentage of the capital outlay spread over a number of years, and he considers that to be the only way that a practical and fair comparison can be made."

CHAPTER X.

STONE LAYING.

THE process of breaking having been decided upon and the stone ready for use, the next work is to apply the material in the way that will give the best wearing and financial results. In this, like in most things, there are different opinions. Some engineers follow Mr. Macadam's method of placing the stones upon the road, and allowing them to bind themselves on being compressed by a roller, whereas others consider that small siftings of the same material should be mixed in with the larger stones, while many others advocate the use of a small quantity of earthy binding material.

Now, when it is known that, with the use of properly broken stones, when spread, nearly 45 per cent. is space, and after being compressed by a steam road roller, interstices still remain to the extent of about 20 per cent., the necessity of using some description of fine material should be clear enough. The binding material employed should not be mixed with the stones until after they are spread over the road, then the binder should be applied in small quantities, not exceeding the interstices in the aggregate.

Should too much binding material be used it is to the detriment of the road, as it prevents the angles of the stones closing together, the result of this being that, as soon as the binder is washed away by rain, the road becomes quickly disintegrated. Care must be taken in instances where the siftings from the stone crusher are used that this material is not placed over the surface in patches, but thoroughly worked into the interstices by means of brooms and water. Should the siftings be allowed to remain in patches on the surface such weak spots will turn to pit-holes soon after the completion of the road. The author has made many observations on the use of binding materials, and is of opinion that when small siftings from the stone breaker are used alone, as much as 30 per cent. of them have to be reduced to grit and mud by rolling before the road surface is properly consolidated. In so doing the cost of rolling is not only increased, but the excess of rolling is also prejudicial to the wearing qualities of the road stone.

One instance amongst many which came before his notice was when some repairs were carried out on a steep gradient with siftings for binding. After a fortnight's wear, during which time there was dry summer weather, the surface was quite broken up, and it was necessary to again repair the road. On this occasion, after some new material had been laid, a small quantity of road sweepings was employed for binding. The roller was employed only two-thirds of the time it took on the previous occasion to properly consolidate the repairs, and when done it stood many weeks of dry summer weather without breaking up. Where earthy materials are employed for binding, the stones, upon being rolled, close together better than when harder substances are used. Any excess of earthy binder, when wet, readily squeezes out from between the stones at the time of rolling, to be afterwards swept off from the surface; but, in the case of hard substances being used to excess, these will prevent the surfaces of the stones from closing up and being in proper contact one with the other.

Mr. Deacon carried out some interesting experiments on the effect of binding material at Liverpool, which were as follows:—"Under a 15-ton steam roller, preceded by a watering cart, 1200 yards of trap-rock macadam, without binding, can only be moderately consolidated by twenty-seven hours' continuous rolling. If the trap-rock chippings from the stone-breaker are used for binding, the same area may be moderately consolidated by the same roller in eighteen hours. If silicious gravel from ¾in. to the size of a pin's head, mixed with about one-fourth part of macadam sweepings obtained in wet weather, be used, the area may be thoroughly consolidated in nine hours. Macadam laid according to the last method wears better than that laid by the second, and that laid by the second much better than that laid by the first."

The following experiment was tried at New York by Mr. W. H. Grant, in consolidating by rolling and compacting road stone by a strict adherence to Macadam's theory. He found that "the bottom layer of stone was sufficiently compacted in this way to form and retain, under the action of the rollers (after the compression had reached its practical limit), an even and regular surface; but the top layer, with the use of the heavy roller loaded to its greatest capacity, it was found impracticable to solidify and reduce to such a surface as would prevent the stones from loosening and being displaced by the action of wagon wheels and horses' feet." He further says: "The rolling was persisted in with the roller adjusted to different weights up to the maximum load (12 tons), until it was apparent that the opposite effect from that intended was being produced. The stones became rounded by the excessive attrition they were subjected to, their more angular parts wearing away, and the weaker and smaller ones being crushed."

CHAPTER XI

ROLLING.

In making a good road the roller is an indispensable article. Its employment creates a considerable saving in the wear and tear of horses and vehicles, besides a large amount of economy by increasing the actual durability of the road. The author recently had cause to make some observations on the latter question effected in the following manner:—A quantity of surfacing material which had been worn in by the traffic, and a similar amount of material that had been rolled in by steam roller, were procured from the same road, each of which had been laid in the same month and year, and subjected to practically the same amount of traffic; 28 lb. dry weight of each sample was then taken separately and washed in a sieve containing sixty four meshes per square inch. The material that remained in the sieve was then weighed in each instance, with the result that there were 17½ lb. of that rolled in by the traffic, and 22 lb. 3 oz. of that rolled in by a 10-ton steam roller. This result is only what might be expected, when it is considered that in the one case the stone is thrown loosely upon the road, to be subjected to the most severe abrasion by the traffic for a considerable time before it is consolidated, whilst in the road consolidated by a steam roller the new stones are at once bedded, with the addition of as little yielding material as possible, and form a hard surface to effectually resist the weight of the traffic and the atmospheric changes. Before a road becomes consolidated by the action of the traffic there must be an undue quantity of pulverised material which in the winter causes excess of mud, necessitating extra cost for cleansing and scraping. Stone which is loosely spread on the road to be rolled in by the traffic should be looked upon as an obstacle to modern traffic and a cruelty to horses. In these days the public expect as much comfort and convenience as the best method of road-making can provide, and this can only be effected by the employment of the steam road roller on the principal country as well as town roads.

Steam road rollers are locomotives with broad heavy wheels or rolls. They are usually constructed to weigh from 10 tons to 15 tons each, but a few are employed as light as 8 tons and as heavy as 25 tons. Ten tons is the most desirable weight for general use, as the water pipes, &c., existing in the roads are not

Fig. 28.

so liable to become damaged under this as would be the case in the employment of heavier rollers.

By the use of a 10-ton steam road roller about 1500 yards, 4in thick, can be consolidated in one day's work at a cost of about one-sixth of a penny per yard super. Horse rollers are not to be recommended for general use on roads. They are more expensive to use than steam rollers, the cost of horse work being great compared with the amount of work done. From two to four horses are necessary to draw the roller, and it is found that where insufficient horses are used, their feet displace a large portion of the work executed. Rolling should always be commenced at the sides of the road, and continued there until the materials become fairly consolidated, after which the rolling should be gradually worked up to the centre. Unless proper attention is paid to this procedure the weight of the roller will press the crown stones down on to the sides. With a 10-ton roller, a greater thickness than 4in. of materials should not be spread on to the road at one time, or the result will not be very satisfactory. The rolling should be done slowly, which practice will not only conduce to more work, but will reduce the cost of fuel, and the stone will not be so much injured as would be the case if a more rapid motion were employed.

The following description of the manner in which road-rolling should be executed is taken from Messrs. Aveling and Porter's booklet on "Steam Road-rolling":—"In the best practice the roadway is excavated, graded, and properly formed to a depth of 14in. from the level of the gutters, with a cross section conforming to the cross section of the road when finished; it is then thoroughly and repeatedly rolled with the steam roller, all depressions being carefully filled and rolled before the stone is put on. On the bed thus formed and consolidated a layer of stones 8in. thick is set by hand, and rammed or settled to place by sledge hammers, all irregularities of surface being broken off and the interstices wedged with pieces of stone. The intermediate layer of broken stone, of a size not exceeding 3in. in diameter, is then evenly spread to a depth of 4in., and thoroughly rolled, and this is followed by rolling in ½in. of sand. The surface layer of stones, broken to a size not larger than 2in. diameter, and to a form as nearly cubical as possible, is then put on to a depth of 3in., thoroughly rolled, and followed as before by sand, also rolled. Finally, a binding, composed of clean, sharp sand, is then applied, well watered, and most thoroughly rolled with the steam roller until the surface becomes firm, compact, and smooth, the superfluous binding material being swept off and removed."

The following table contains a summary of the road-rolling work done on the County Council roads of Nottinghamshire, contained in the annual report of the County Surveyor for the year 1891-92.

	Without team labour for watering.				With team labour for watering.		
	£	s.	d.		£	s.	d.
Total cost of steam rolling	459	5	7½	...	656	6	6½
Averages:							
23,444 tons 4 cwt.—being the quantity of material used during the year—per ton ...	0	0	4¾	...	0	0	6¾
303 miles—being the mileage of the county—per mile	1	10	3¼	...	2	3	2¼
252 days to the year—which is about the number of days the rollers actually work—for one roller, per day	0	12	1¾	...	0	17	4

In the annual report of the County Surveyor of Monmouthshire, 1898, it is shown that the cost of road-rolling 14,950 tons of material worked out at 6¾d. per ton of material rolled, which includes manual and team labour.

The illustration (Fig. 28) is one of Messrs. Aveling and Porter's steam road rollers. This firm of engineers were the pioneers of steam road roller manufacturing, and their high quality of workmanship, together with over thirty years' experience, enables them to retain their position in the front as successful manufacturers. All the foregoing figures on steam road-rolling are from work done by Messrs. Aveling and Porter's rollers. Steam road rollers, when not engaged in rolling, can be turned to other profitable uses. They can be employed to drive the machinery for breaking stones, for which purpose the author has on several occasions availed himself of their use. Rollers can, further, be employed to work fixed machinery; or, if provided with a set of extra wheels, are suitable for traction work. Fig. 29 illustrates an Aveling and Porter steam roller attached to a stone-breaking machine.

Watering is an important adjunct to road-rolling; it expedites the process, and reduces to a large extent the abrasion and crushing of the metal that would otherwise take place.

The ordinary watering carts used for street watering should not be employed with a steam roller, owing to their wheels being too narrow. These narrow wheels disarrange the stones by being pressed by their heavy load in amongst them, which further entails a much heavier load for horses to draw, causing their feet to displace the road materials. It is much better to employ watering carts provided with good broad wheels in this work. A hose pipe,

Fig. 29.

attached to street fire hydrants, is sometimes employed instead of the watering cart. Whatever method is adopted, care must be taken that too much water is not put on the road, especially at its early stages, or the foundation may become soft and weak; and, further, the water should never be applied in bulk, but sprinkled on through a rose distributor.

CHAPTER XII.

ROAD-BREAKING.

In repairing roads it is desirable to loosen the surfacing stratum of old materials by some means or other before laying on new stones. By so doing a large quantity of the old material may be utilised over again, and the road may be more easily re-formed. This process, further, affords a more compressible bed for the new road material. The method usually adopted is by hand picking, in which process the road surface is broken across at intervals of from 5in. to 8in. Grading picks are employed in this work, which are made of wrought iron, pointed at each end with steel, and generally weigh from 6 lb. to 9 lb. They are each provided with an eye in the middle, in which is fitted an ash handle of 2ft. 9in. in length. The cost of raising a road in this manner is on the average about 1½d. per yard super. It is, however, desirable that the whole of the surface should be broken up regularly, which enables a better wearing surface to be made at the finish.

It is not always desirable to lift a road before laying on the new material, in cases where the superficial stratum is worn very thin, and a coating of 3in. or more is necessary; material can often be applied with advantage without breaking the road. The surface should not be broken through to the road bed, as in so doing the foundation of the road is sometimes interfered with, and softened by watering or wet weather.

Of recent years many contrivances have been introduced to perform the work of road-breaking by machinery, some of which have been a success. Spikes inserted at intervals round one of the driving wheels of a steam roller have been employed to perform this duty, but this is a method not altogether to be recommended. More recently a less costly process has been introduced in that of the scarifier, which, upon being attached to a steam roller, performs its work both rapidly and economically. The road scarifier picks up the whole or any portion of a macadamised road surface where required, and, as a rule, it will give good results on surfaces metalled with hard granite.

The author has witnessed one of Messrs. Aveling and Porter's "Morrison Scarifiers" break up some of the hardest granite roadways with little trouble. This excellent contrivance is illustrated in Fig. 30. It can be worked either backward or forward, and quite near to the channel or kerb. It is attached to the roller, so that at any moment it can be put into action. The first cost of this scarifier is £94 net, at Rochester, and that of

Fig. 30.

fixing varies according to the facilities at hand for doing the work, but does not as a rule exceed £25.

There are other makers of scarifiers beside Messrs. Aveling and Porter. Rutty's scarifier is one of the best appliances in use. It consists of a heavy cast iron frame, into which is fitted six movable steel tines, or teeth, three at each end; one set to be used at a time, according to the direction in which the engine is moving. The machine is attached to an ordinary steam roller by means of a connecting chain, which relieves the engine of much vibration which otherwise tends to shake the machinery. This scarifier is simplicity in itself, and is a capital worker. The cost of this machine complete, including the services of a competent man for two or three days, till the driver is well acquainted with

Cost of Breaking up Road by Rutty's Scarifier.

Portion of Great North-road.	Time. Hours.	Wages.	Horse hire, water, fuel, and oil.	Wear and tear of scarifier.	Wear and tear of engine.	Total cost.	Area. Super yards.	Total cost per super yard.
		£ s. d.	£ s. d.	£ s. d.	£ s. d.	£ s. d.		d.
Finchley Park to Woodside-lane	9	0 13 1	1 2 0	0 4 6	0 18 0	2 17 7	1320	·523
Hertford-road to Church-lane	8½	0 11 6½	1 1 3½	0 4 3	0 17 0	2 14 1	1066	·609
Cherry Tree Hill	10	0 13 7	1 6 0	0 5 0	1 0 0	3 4 7	1950	·397
Church-lane to Lee House	5	0 6 9½	0 11 9	0 2 6	0 10 0	1 11 0½	670	·555
Police-station to Tottcridge-lane	7½	0 10 3	0 18 7	0 3 9	0 15 0	2 7 7	1030	·554
Friern-lane to Old Swan	5	0 8 11	0 13 9	0 2 6	0 10 0	1 15 2	857	·493
				Total average ...		14 10 0½	6893	·504

the use of the scarifier, is £105 net. The cost of the employment of this scarifier has been carefully tabulated by Mr. Francis Smythe, C.E., for the information of the Finchley Council, and the foregoing is a copy of his figures.

Fig. 31 illustrates a new scarifier, manufactured by Messrs. W. Thackray and Son, Malton. This machine is not a fixture to the engine, but can be easily attached or detached. It has four teeth, in order that a greater width may be picked up at a time than is generally the rule.

This machine is simple in construction, and no doubt has a good future before it. Its cost is stated to be £60 complete.

The following are some opinions of municipal engineers on the employment of the scarifier, to whom the author is indebted for the information :—

Dundee.—A scarifier would be preferable, and a saving would be effected by its adoption. Have recently experimented with scarifier in excavating track for tramways in macadamised road, and found a saving in labour of at least 60 per cent.

Portsmouth.—The saving is, I estimate, 33 per cent. in favour of scarifier.

Swansea.—I consider that scarifying costs about half as much as hand labour.

Carlisle.—A scarifier does three times the work of hand labour at a given cost.

Great Yarmouth.—The roads are much better broken up with a scarifier than by hand labour at one-tenth cost.

Wells.—I have tried the road scarifier, but find the cost is so very much in excess of hand labour that I have abandoned its use.

Brighton.—I find the saving about 66 per cent. over that of hand broken.

Derby.—Great saving compared with hand power.

Llanelly.—Have used a scarifier, which is a saving of at least 50 per cent., taking everything into consideration.

Eastham.—About half as costly as stocking by manual labour.

Tiverton.—Have only just begun to use a scarifier, but there is an enormous saving in labour and materials.

Hanley.—Half the cost saved.

Fig. 31.

Canterbury.—The saving effected by using a scarifier as against hand labour is enormous.

Aberdeen.—The scarifier with three men will do as much in one day as ten men will do in a week by hand.

CHAPTER XIII.

PAVED ROADWAYS.—WOOD PAVEMENTS.

THE use of wood for street paving purposes is growing more popular every year, and is not to be wondered at, when the many advantages of its employment over most other materials are considered. Wood pavements have been adopted largely in this country with much success, increasing comfort and decreasing the expense of all concerned. The advantage of wood over granite or asphalt for pavements, regarding comfort and wear and tear of horses, are considerable, there being a less amount of jar to the horse, together with a more secure foothold.

Since the time wood pavements were first introduced many endeavours have been made to provide a material that would exceed all others in comfort and durability. This has been attempted in various ways, and the following are some of the methods employed:—In order to increase the life of various woods, different chemicals have been used to retard decay, by counteracting those influences which are conducive to this.

Creosoting consists in subjecting the wood to immersion in the oil known as creosote, which soaks or is forced into the pores of the wood, serving both to keep out the atmospheric influences and to check decomposition. A method often resorted to in order to impregnate the wood to as large a degree as possible with this oil is by pressure; the timber is placed in an iron boiler containing boiling creosote oil, the boiler is then closed, and an internal pressure of 130 lb. to the square inch is applied, which forces the oil into the wood. The wood is removed from the boiler when it has absorbed from 8 lb. to 10 lb. per cubic foot. Creosoting sufficient for ordinary purposes may be effected by placing the wood blocks in the boiling liquid prepared in an ordinary tar boiler. After ten minutes' immersion they will be ready for use. It appears that blocks so treated wear equally as well as those which have been treated under great pressures. When blocks of soft wood are treated under great pressures to force the preservative substance as nearly through the wood as possible, the life of the wood is more often than not reduced, through the grain and pores of the wood being injured; at any rate, the resulting wear will not warrant the extra cost incurred. Great care must be taken that the wood before being creosoted is perfectly seasoned and as free as possible from sap, otherwise this process, which is applied in order to prevent decay, will cause rapid fermentation to take place in the heart of the wood, and will within a comparatively short period produce rot and rapid deterioration. The same precaution should be taken before any preservative course is applied; attempts to season wood after the process is finished are quite useless.

Burnett's Process.—In this process the wood is impregnated with a solution of chloride of zinc and water in the proportion of 1 of chloride of zinc to 45 of water. This process is generally performed under pressure of 130 lb. to the square inch, but can also be effected by immersing the blocks in the solution for a lengthy time. This process renders the wood uninflammable.

Gardener's Process.—By this process it is intended to dissolve the sap and drive out the moisture by the use of chemicals, leaving the fibre only to remain. Further chemicals are employed in an after process to both render the wood more durable and incombustible. This process generally takes a week or ten days to perform.

Ryan's Process.—The timber in this process is immersed in a solution of corrosive sublimate or bichloride of mercury and water, in the proportion of 1 of the former to 130 to 150 gallons of water, according to the strength required. The period of immersion is about 30 hours per inch in thickness for soft wood paving blocks, increasing with the density of the wood. This process is valuable in retarding dry rot.

Wood pavements are usually constructed of blocks measuring from 8in. to 10in. in length, 3in. to 5in. in width, and from 5in. to 8in. in depth. They are laid grain up, or, in other words, with the fibres of the wood in a vertical position to the traffic, which is the best position both for wear and for affording a foothold for horses. Some wood is better when used in the whole section of the tree, with the bark removed, but without the wood being cut into shaped blocks. This is the case with cedar, which is largely employed in other countries for pavement purposes. When cedar is cut up into the usual rectangular paving blocks it is found to be liable to split through the grain, which is not the case when it is used in the whole section.

Some of the best-known systems of wood pavements are :—

Asphalt Wood Paving consists of a concrete foundation covered with a layer of mastic asphalt ½in. in thickness, on which creosoted wood pavement blocks are placed with about ½in. spaces between each row. These spaces or joints are then partly run in with melted asphalt, which readily unites with the similar material upon which the pavement is laid. The remaining portion of the joints are then filled in with cement grout or lias lime.

Carey's Wood Pavement consists of paving blocks shaped with interlocking ends. It is claimed for this pavement that no special foundation other than a layer of fine gravel over the ordinary road bed is required.

Harrison's Wood Paving differs from the asphalt wood paving only in that, instead of the under layer of mastic asphalt, the asphalt material is run into the joints and under the paving in one process. This is effected by the blocks in the first instance being raised about ½in. from the foundation on strips of wood, so that when the hot asphalt is run into the joints it flows under and fills up the spaces between the foundation and the pavement.

Henson's Wood Pavement.—In this system the paving blocks are laid on a concrete foundation, covered with well-tarred roofing felt, a strip of which is also placed between every other course of blocks whilst being laid. The blocks are driven as closely as possible together on completion of every eight or ten courses with heavy sledge hammers, a plank being previously placed against the face of the wood. On completion the whole is grouted in with pitch. The introduction of felt in the manner adopted is intended to give elasticity to the road.

The Improved Wood Pavement.—In the system originally employed by the Improved Wood Paving Company wood blocks previously dipped in tar were laid on two thicknesses of 1in. tarred boards, laid so as to cross over each other at right angles, and which were placed on the ordinary road bed.

The system more recently adopted consists of the following:—Upon a foundation constructed of a bed of 6in. Portland cement concrete, faced over with $\frac{3}{4}$in. thickness of cement mortar to the convexity required for the pavement when finished, a covering of creosoted red northern fir blocks—each measuring $8\frac{3}{4}$in. long, $3\frac{1}{8}$in. wide, and 6in. deep—is constructed. The blocks are set with close longitudinal joints, and with $\frac{3}{8}$in. transverse joints. The spaces between the blocks are regulated by the temporary insertion of strips of wood, which, upon the pavement being laid, are removed from the joints. Hot pitch is then run into the open spaces, which fills any irregularities that may exist between the foundation and the wood covering, besides partially sealing the joint. The remaining portion of the joints is filled with a grouting of neat cement, broomed over the surface until it is rendered as impermeable as possible. Previous to the traffic being admitted on to the new work, a coating of fine gravel is spread over the surface, which works in by the wheels and forms a hard crust to the face of the pavement.

Hard Woods.—By far the most important of all wood for pavement purposes are those known as Jarrah and Karri. An instructive article appeared in the columns of *The Illustrated Carpenter and Builder* on the question of wood for street pavements, from which the following is quoted:—

"In 1884 London counted—according to the statements of M. Petsche, for many years the municipal engineer of the City of Paris—eighty-five kiloms. of wood-paved roads, twenty-one kiloms. of asphalt roads, 923 kiloms. of macadam, 1269 kiloms. of boulders, and 451 kiloms. of granite. But already the arterial thoroughfares of the metropolis were paved with wood to the extent of 820,000 square metres. Baltic deals were then—at the time of the South Kensington Exhibition—ascendant for the railways; and surveyors had to be convinced of the benefit of a change from soft to hard wood. There were some active, clear-minded men on the vestry boards and in the works departments, and the result was that a move was made; and in three years a sample paving with Jarrahdale jarrah blocks was made over a considerable section of the Westminster Bridge-road. In London, in 1839, the first wood paving was laid in the Old Bailey, and in 1889—half a century later—the grand change from softwood to hardwood was initiated.

All the early-dated softwood paving has long since been worn out and removed, and the existing softwood pavements subsequently laid are fast being rooted up. But the first piece of West Australian Jarrahdale laid in Westminster Bridge-road remains in use—good, firm, and serviceable, although the heavy and continuous traffic over it is not surpassed by any other heavy traffic route in the whole metropolis. This work was done with the objectionable wide joint, and, consequently, has not had a fair chance. In an official return of the Lambeth Vestry, made by request of the Paddington Vestry, this extraordinary endurance is thus briefly noticed:—Vestry—Lambeth.—Description of wood preferred—'jarrah.' Size of blocks, 9in. by 3in. by 5in. (depth), open $\frac{3}{16}$in. joints; grouting in cement. How long has it been laid? Six years (to October 24th, 1895). Depth of wear, $1\frac{1}{16}$in. in six years at Westminster Bridge-road. Cost per 1000 blocks, £8 16s. Noise.—'Do you consider the hardwood better than Swedish deal as to noise?'—'No.' 'As to slipperiness?'—'No.' 'As to shrinkage?'—'Yes.' 'Does it require more scavenging than Swedish deal?'—'Less scavenging.' 'Is it more slippery in frosty weather than Swedish deal? 'No.' General remarks.—The Surveyor of Lambeth says, 'There is no comparison between deal and hardwood as to durability and quality of road and cheapness, although the first cost of hardwood is greater.'

"In the early days of wood paving it was the immediate cheapness of a wood in common use for building purposes that caused the introduction of Swedish deals. Their deficiencies in wearing life; the insanitary conditions to which they were reduced in their applications to paving by the organic refuse of the streets; their shrinkages and expansions in varying weather; their wet rot and dry rot; and their numerous other irremediable defects inevitably suggested—it may, indeed, be said to have enforced—that competition by the hardwoods which has now become so masterful.

"The matter of price further held the minds of vestries and municipalities, although here and there small sections of some of the well-known timbers—oak, teak, beech, chesnut, elm, and ash—were tried, as also were some foreign trees. The supply, however, seemed limited. It was not so with the forests of Western Australia, which presented almost inexhaustible resources. The embarrassment was then to choose from the midst of numerous varieties of species that which would afford the most desirable quality at an acceptable price.

"This difficulty has been solved by the Jarrahdale jarrah, said to be the finest road-paving ligneous material in the world. And regarding the forest holding of the company which owns it—and which is more than half the total of the acres leased by the Western Australian Government—250,000 out of 447,000, or nearly equal to the combined holdings of all the other timber leases in the Colony, it should have ample resources for maintaining its best commercial position in the open competition of purchasers.

"Of the superior economy of the Jarrahdale jarrah there is no doubt, and the rough formula, twice the cost of deals and three times the life. As the Paddington Vestry's special committee put it in their report:—'Sixty per cent. longer life is required for cost of hardwood to be equal to the cost of softwood.'"

"Jarrah," says Mr. Ednie Brown in his recent report upon "The Forests of Western Australia and their Development," "is without doubt the principal timber in the Western Australian forests; no one knowing the subject would for a moment dream of classifying it as anything else. It is predominant above all others in its extent of forest, the various uses to which it is or can be applied, the part which it is now taking in the great timber export of the Colony, and the esteem in which it is held in the country. Jarrah and Western Australia are almost synonymous words, and, as this has been the case from the earliest days of the foundation of the Colony, so it will now remain as long as a Jarrah forest exists. I do not mean by these remarks to disparage in the least degree any of the other commercial woods of the country, but simply to emphasise the fact that Jarrah is the principal indigenous timber of this part of the Australian continent. There are other timbers in our forests which are equally, if not more, valuable, for their own special purposes, as I shall point out; but for general constructive works necessitating contact with soil and water, the timber of this tree stands foremost in these forests. . . .

"Its resistance to white ants is remarkable, and houses built of the wood when thoroughly seasoned are almost indestructible, and have been known to exist in perfect preservation for nearly one hundred years. It gets extremely hard with age, and then becomes almost unworkable; even strong nails cannot be driven into it, and when struck the wood rings like a bell. Altogether it is a remarkable timber, and is highly suited for outside works. Should any decay or destruction have occurred in the timber after having been years in use, it would always be found that this is confined to the sap-wood, which, therefore, ought always to be avoided in the construction of houses or in other works of a permanent character."

The following is from a work by the late Baron von Mueller, of Melbourne :—

"***Eucalyptus Marginata.***—The Jarrah of South-Western Australia, famed for its indestructible wood, which is attacked neither by the chelura or teredo or termites, and therefore so much sought for jetties and other structures exposed to sea water; also for underground work, and largely exported for railway sleepers. Vessels built of this timber have been enabled to do away with all copper plating. It is very strong, of a close grain, and a slightly oily, resinous nature; it works well, makes a fine finish, and is by Melbourne shipbuilders considered superior to either oak, teak, or any other woods."

Mr. Ednie Brown states "that there are many cases on record where Jarrah piles, drawn from bridges and jetties, after being over fifty years in sea water swarming with the teredo, have been found to be almost as sound as when first driven."

Karri, says Mr. Brown, "is not so well known as the Jarrah, owing to the limited field of its growth and the, at present, comparative inaccessibility of its haunts." He describes the bark as "smooth, yellow-white in appearance, but not persistent, like the Jarrah. It, therefore, peels off in flakes each year, and thus the tree has always a clean, bright appearance. In consequence of this it is frequently spoken of as a 'white gum,' although generally known as the Karri."

He further states that the "timber is red in colour, and has very much the appearance of Jarrah; indeed so like are the two that it takes a good judge of both to distinguish them. It is hard, heavy, elastic, and tough, but does not dress, nor can it be wrought so easily as its contemporary." The best means of discriminating betwixt the two timbers is to ascertain by burning and comparing the ash left. Jarrah, whilst burning, looks black, leaving practically no ash, while Karri, being burnt, leaves a good deal of white ash, similar to red gum. Mr. Brown considers Karri to be equal to, if not better than, Jarrah for street pavement purposes, owing to the fact "that the wear caused by the traffic does not render it so slippery for the horses' feet." The life of these woods for paving purposes has not been positively arrived at, as their introduction has been of an insufficient period to determine this question. However, when Jarrah and Karri are employed they are each expected to have a duration of fourteen to sixteen years.

CHAPTER XIII.

PAVED ROADWAYS.

JARRAH and Karri blocks for pavements do not expand to any great extent when exposed to moisture, their relative expansion being much less than that of soft woods. In cases where insufficient attention has been paid to seasoning the timber before use, trouble generally ensues from excess cf contraction or expansion. Too much attention cannot be given to the question of seasoning—unseasoned blocks will shrink in hot weather and expand when exposed to moisture; the result of the former being that the joints become open, allowing matters to soak away between them, perhaps to the danger of the foundation; they further render the pavement unsanitary by allowing noxious matters to sink below the surface instead of running off or being removed. The joints thus being opened, soon become filled up, or are regrouted in cement, but the result of this is that in the wet season the wood blocks swell, and being unable to re-occupy their former space, need provision to be made to allow for this expansion.

Even when well seasoned materials are used it is necessary to make provision for expansion. This should be effected in the channel courses, the blocks of which are turned with their lengths along the street. Sometimes two of these courses are temporarily provided with 1in. joints filled in with sand or soft mortar. Sometimes one 1½in. joint is left next to the kerb filled in with sand; this is not to be recommended. It is much better to leave two joints of less width, with the one further from the kerb filled in with a clay substance, and the other nearer or next the kerb filled in with mortar. In the case of a porous joint being provided next the kerb, a greater quantity of liquid and unsanitary substances soak into it, to the detriment of the pavement, than would be the case if the joint were made with clay and placed a few inches further from the kerb. Fig. 32 shows a plan and section of a wood-paved road.

Where perfectly-seasoned Jarrah and Karri woods are employed, the expansion is but a trifle. The expansion that took place in a 24ft. road that the author laid two years ago with Jarrah has amounted altogether to 1in. in 10ft. in width.

Three Jarrah blocks which had been seasoned for some months were, after being measured, placed before the fire for twelve hours, and upon being re-measured, their size had not altered. One

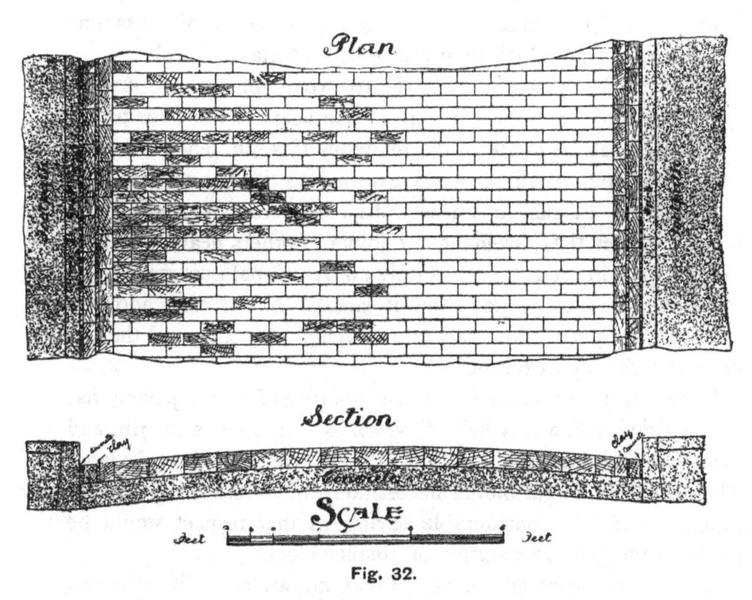

Fig. 32.

block, however, had slightly warped. These blocks the author then weighed, afterwards placing them under water, where they remained for thirty-six hours, and upon being re-weighed, there was an average difference of 1 oz. 12 drachms per block. There was, however, no percepible alteration in the size of any of them by measurement, and upon the blocks being snapped through the centre, the greatest depth to which the water had penetrated the grain of the blocks was $\frac{1}{16}$in., and the least depth $\frac{1}{32}$in.

Similar experiments to the above were performed with Karri wood. From the author's observations, when Karri blocks, originally measuring 9in. long and 3in. wide, are reduced by seasoning to $8\frac{3}{4}$in. long and $2\frac{7}{8}$in. wide, they are then about ready for laying. Some of these blocks, which were laid prematurely, recently came under the author's notice, and the result was not altogether satisfactory. The comparative amount of absorption that takes place in the two woods is difficult to arrive at, since so much depends on seasoning and other conditions. The author found the average absorption of each of the three perfectly-seasoned Jarrah blocks previous referred to, to be less than that of unseasoned Karri by 4 drachms of water, whilst Karri wood, after being perfectly seasoned, gave a conflicting result of 2 drachms less absorption per block than the previous tests of Jarrah.

Size of Jarrah and Karri.—Hard wood blocks are usually specified to be 9in. by 3in. only. All users of wood for paving purposes must be aware of the tendency this 9in. length of block has to warp out of shape when exposed to the hot weather soon after laying. Blocks so warped often become loose, and, unless removed from the pavement, the joints of others nearest to them also become injured. This is only an occasional occurrence, but would be practically prevented if smaller blocks were adopted, such as 7in. by 3in., 6in. by 3in., or even 5in. by 3in., and there is no reason for their exclusion.

In the Eastern Colonies a large quantity of wood paving has been carried out, nearly half of which is with blocks of 7in. and 8in. long, and 6in. by 3in. blocks are much used in New Zealand. The use of the larger blocks necessitates much waste of valuable timber, so that a considerable saving in that respect would be made by the general adoption of smaller blocks.

About two years ago, a new wood, known as Californian red gum, was introduced to the notice of the public by Mr. Edward Alcott, and has since been tested in many parts of London, with, it appears, exceedingly good results.

The valuable information recently given in the columns of the *Surveyor and Municipal and County Engineer* on the trials of this wood in the busy London streets is here quoted as an excellent description of its wearing and sanitary qualities:—"It is claimed that the wood forms a medium between soft woods and the very hard woods that have been laid down in London for some years past; but that, while being neither so soft as the one nor so hard as the other, it combines the advantages of both, and is at the same time more durable than either. The wood is claimed to be non-porous, non-slippery, and as free from noise as it is possible for any wood paving to be. The natural life of the wood, which is used without creosote, oils, or any other preservative, is stated to be from fifteen to twenty years. It cannot be denied that the appearance of the wood is certainly very much in its favour. When closely examined, even through a strong magnifying glass, it seems to be as free from grain and fibre as it is possible for a wood to be, and, whether the result be due to the natural quality of the wood or to the mode of treatment before being cut down, the fact remains that when a block of the wood is cut with a sharp knife the surface presented has in appearance the uniform consistency and smoothness of leather without its softness. It is asserted that the peculiar quality of the wood is such that it cannot tear at the edges. On the contrary, it is claimed that the wood possesses a serviceable degree of malleability, which results after a time in the blocks being so welded together as almost to form, to the eye at least, practically a single piece with a uniform smooth surface. The advantage of this, it is claimed, is that there is none of the surface inequality which is found in the case of so many streets after the paving has been down for a certain length of time, and consequently, none of the jolting and discomfort of which passengers so often and so justly complain. The water being unable to penetrate the joints, there is no chance of pools forming beneath the blocks, and the latter consequently sinking.

"The non-porous character of the wood arises from its density; and here more particularly comes in the sanitary aspect of the question. If the wood resists moisture, both at the surface and at the joints, it does not seem possible that it should be able to absorb any dangerous germs, and only reasonable attention in the way of cleansing should be required to constitute a thoroughly sanitary paving. As to durability, we have been informed that portions of the wood have been in the ground for twenty years, and have then shown no signs of decay or effects from damp, and that railway sleepers laid for over twelve years in virgin soil are as sound as when first put down. Again, although presenting a sufficiently smooth surface to give a minimum of friction to traffic, it is claimed that the wood, when laid, is not slippery, presents a thoroughly good foothold for horses, requires no gravel or grit, and that the surface is therefore more free from dust and more sanitary than in the case of other wood pavements. A quality upon which special emphasis is laid is that the non-absorbent quality of the wood necessarily implies that it is not subject to expansion from moisture. From the information we give below it will be seen that this claim has survived a somewhat severe test. We have thus endeavoured to state briefly and concisely the claims put forward on behalf of the new competitor. The subject of a thoroughly satisfactory paving material, whether wood or otherwise, is one in which our readers are deeply interested, and it occurred to us that it would be acceptable to municipal engineers and others if we made some independent inquiries as to the extent to which the wood has been actually tried in London, and what the results are so far as can at present be judged.

"So far as street paving in London is concerned, the new wood has probably been as severely tested in Westminster as anywhere. It was brought to the notice of Mr. Wheeler, the surveyor to the vestry, two years ago, and he decided to test it, as he has done in the case of other materials which seemed likely to justify the time, trouble, and expense. For about a week he soaked a block of the wood in water, and had found the increase in weight to be little more than 3 oz. He also placed a block in front of a fire for a whole day, and found that there was practically no change, having previously traced its dimensions on a piece of paper. Having thus satisfied himself as to the character of the wood, he recommended to his committee that they should make a trial of it, and a crossing was laid in Bridge-street from Parliament-street. On the one side of the Californian red gum were laid Australian hardwood blocks, and on the other dipped deal blocks. The kerbs abutting on the jarrah blocks have been re-laid six times during the twelve months the wood has been down; in the case of the new wood it has not been found necessary to relay or alter the kerbs at all. It may also be pointed out that not only are the blocks laid from kerb to kerb—in itself a sufficiently good test—but that there is a refuge in the middle of the roadway. If the wood was subject to expansion it would inevitably disturb the kerbs.

So satisfied is Mr. Wheeler with the results of this test that he has decided to re-lay a large portion of Grosvenor-road—about 3000 yards—with Californian red gum, and also to repair the crown of Whitehall with it. Its superiority over other woods, in Mr. Wheeler's opinion, consists in the fact that it forms an even, non-slippery surface, causes less dirt, as owing to its pliability it does not tear off like soft wood or break off like hard wood. Although Mr. Wheeler has laid about 115,000 square yards of all kinds of wood, he considers, from the peculiar formation of this wood, that it is the most suitable material for London roadways that he has seen. It is non-absorbent, and on sanitary grounds it is good. Mr. Wheeler has taken up some of the blocks first laid down, and has found that absolutely nothing has penetrated them."

Amongst soft woods Swedish yellow deal and Baltic fir are considered to be the two most valuable for wood pavements, and have been very extensively employed in London and elsewhere; yellow pine and beech have also been largely used. The life of any one of these timbers is considered to be from seven to nine years, according to traffic. A disadvantage to the employment of soft woods is that, owing to their power of absorption, the surface of the road remains wet for a long time after rain, which is certainly a discomfort to horses, they being apt to slip on the wet surface of wood, unless it is constantly gravelled. Hard woods dry much more quickly. It should further be remembered that the wood that absorbs the least quantity of moisture is the one to be preferred from a sanitary point of view, a feature of most important consideration.

CHAPTER XIV.

JOINTING AND FOUNDATIONS.

CLOSE blocking is to be recommended in preference to the system of jointing in all descriptions of wood pavements. Wide joints have been adopted to a great extent, but the system is fast dying out. Such pavements, after a short period of wear, become quite uncomfortable to travel over, owing to the jointing material wearing down below the surface of the pavement, allowing the arris of each row to be worn or knocked off. The wear of the material itself appears to be also affected by this method; not only do the arrises become damaged, but the fibre of the wood has a greater tendency to spread and disintegrate under the traffic, affording greater facilities for absorption, causing greater amount of expansion, and the material to become saturated to a larger degree with unsanitary matters. Close blocking is carried out by the employment of hot pitch and creosote oil, mixed, or by the use of bitumen.

The material used should be boiled in an ordinary tar boiler, and delivered hot on to the wood in shallow dipping trays. The blocks are then dipped three-quarters of their depth into the boiling contents of these trays, and afterwards laid in their places, one against the other, in perfectly straight lines from kerb to kerb. When half a dozen blocks are in their places, a flat piece of wood is placed against their face, and with the assistance of a wooden mallet the blocks are driven perfectly close together. On completion of each day's work the surface should be well grouted in with hot boiled tar, and afterwards well sanded on the surface A grout of cement and sand in proportion of 1 of cement to 4 of sand is sometimes employed instead of tar and sand. The transverse joints of the rows of blocks should always be broken by commencing every alternate course with a half block. Should this not be done the traffic, in following the unbroken joints, would sooner or later form ruts. At junctions of all roads the blocks should be laid diagonally, as shown in Fig. 33, to effect the same purpose as described. The three or four courses nearest the kerb should be laid lengthways with the street, to form a channel-way for removing the surface water (Fig. 32, Chap. XIII.). It is here that provision is usually made for expansion by allowing wider joints, as already described. These joints are best made by the use of strips of wood of the thickness of the joint required, temporarily placed between the layers, and afterwards removed.

Foundations.—The importance of good foundations for wood pavements cannot be over-estimated. Wood paving once laid is

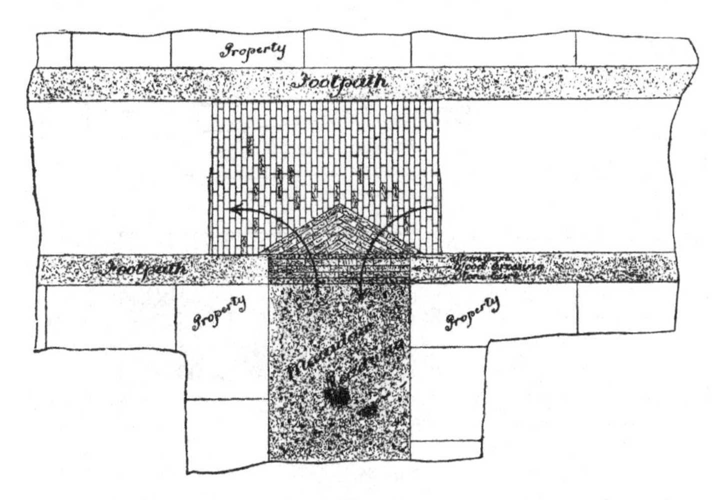

Fig. 33.

expected to remain for many years before renewal; but unless the foundation is quite sound, that will be the first to become defective, and will cause the superficial pavement to disintegrate and decay. Good foundations, when once laid, will not require renewing when the paving is worn out, and the same foundation will serve to carry many renewals of the superficial pavement. The construction of foundations lacking solidity, such as is sometimes practised, by merely placing a layer of sand over the natural road bed, on the score of cheapness, should be guarded against.

Concrete as a foundation, when properly constructed, is not to be beaten. A layer of 6in. in thickness should be placed over a properly-formed bed, the latter having been previously consolidated by a steam road roller. The face of the concrete should be finished with a fairly smooth surface in the form required for the road. This may be accomplished by the use of screeds, which are long pieces of wood generally 1in. thick, which are made by a carpenter in the form required for the road when finished. These screeds are then placed on the road bed at a distance of a few feet apart; the space between them is then filled up with concrete and consolidated by ramming till the

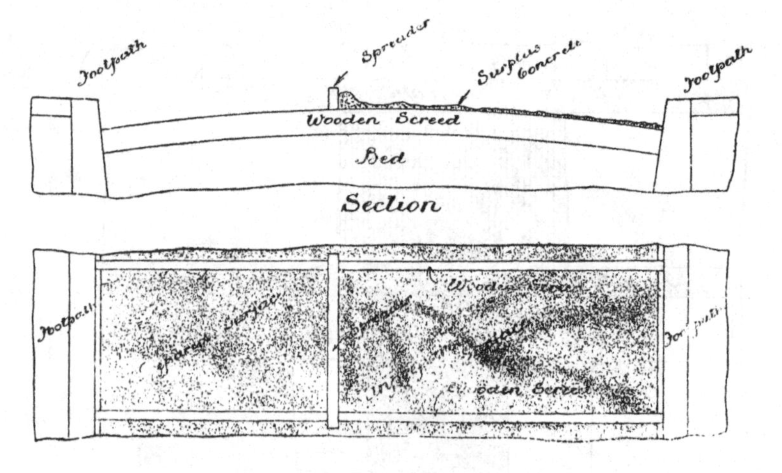

Fig. 34.

material is about ½in. below the top of the screeds; the remaining depth is then filled up with finer concrete, and by means of a straight edge is levelled down fair to the form of the screeds, as shown in Fig. 34.

A cheap and effectual foundation for districts where stone is plentiful is to construct, in the first place, a good Telford foundation, finished on the upper face to the curve required for the paving when complete; over the face of these stones a layer of 2in. of good concrete should be placed. A foundation so constructed is more easily removed for the examination and renewal of under-coursing pipes than is the case of a thick and solid body of concrete. In many cases where good foundations of concrete are laid, a 1in. layer of sand is spread over the surface to act as a cushion for the wood paving. The wood blocks are then set upon this coating of sand and rammed to an even bearing.

This method is not altogether to be recommended, as there is no practical gain in its adoption. As regards its affording a cushion to the paving laid on it, this is inconsistent with the object aimed at, for the bed on which the blocks are laid must be sufficiently substantial to be unyielding. Further than this, as soon as water finds its way through the joints between the blocks the sand will be saturated, and become mobile under the varying weights of the traffic, which is deleterious to the best wearing qualities of the pavement supported by it, besides the likelihood of the sand becoming saturated with impurities. A method far to be recommended in preference to sand is the use of a layer of tarred roofing felt over the concrete similar to that adopted in Henson's pavement.

At the time of constructing foundations for paving purposes, it is most desirable that all water, gas, and other pipes and services, and electrical or other conduits, should be examined and renewed where necessary, and sufficient inspection chambers should be fixed, so that ready access may be obtained without disturbing the pavement.

CHAPTER XV.

WOOD PAVING FROM A SANITARY STANDPOINT.

ONE often hears it said that wood pavements for street purposes are unhealthy, and occasionally very strong objection is taken to their employment on sanitary grounds.

There are, however, many differences of opinion on the point as to the extent, if any, that health is endangered by the use of wood pavements. Some protest against their use on the grounds that the wood absorbs a large amount of impurity, which in the hot season decomposes, and taints the air with unhealthy vapours, whereas others contend that less objection, on sanitary grounds, should be taken to the employment of wood than to many other systems commonly adopted.

These differences of opinion may to a great extent be accounted for by the method of laying the blocks and the materials used. Hard wood paving, well laid with water-tight joints on a good foundation of concrete, are to be preferred, from a sanitary standpoint, to those pavements constructed of macadam or soft wood, while creosoted soft wood may be considered a more sanitary road covering than macadam, the cleansing of wood paving being, as a rule, much less costly than in the case of macadam.

The power of absorption of an efficiently-constructed wood pavement, as compared with that of a well macadamised road, needs no comment; it is only necessary to pour a pint of water on the one and an equal quantity on the other, and while this will nearly all run off the wood, on the other hand, it will nearly all soak away into the macadam. After this compare the length of time it will take for their surfaces to dry in the sun, and it will be found that the wood will take but a fraction of the time as compared with the macadamised road.

Wood pavements can be kept practically free from dust in dry weather, which is not the case with many other road materials.

Dust is always more or less full of impurities, which become wafted by every air current into the atmosphere, invading dwelling-houses, and being inhaled by the human subject. A block of Jarrah wood was recently taken up from a paved road for the author's inspection after nearly two years' wear, and although the week preceding had been one of nearly continuous rainfall, yet upon the block being split the greatest depth to which there were any traces of penetration was $\frac{1}{8}$in. A macadam road, in continuation of this pavement, upon being opened, was found to be wet to a depth of $1\frac{1}{4}$in.

Health statistics of towns using large quantities of wood pavements have been compared with those of towns that have not adopted them, and the results have been rather in favour than against the employment of wood. Colonel Haywood remarks in respect to the healthiness of the employment of wood pavements :— "It has been said that wood pavements at times smell offensively and may be unhealthy, but although some city streets have been paved with wood for thirty years, no complaints that I am aware of have been made to the Commission on this head, and the inhabitants at all times have not only expressed great anxiety lest the wood should be replaced by other materials, but have subscribed toward the cost of its renewal." He continues by saying :— "I have at times noticed offensive emanations from it near cab stands, but am unable to find further evidence of its unhealthiness. These remarks must be held to apply only to public streets open to the sun, air, and traffic ; in confined places and under some conditions wood might be objectionable. I have seen it decaying in confined places without traffic."

Regarding hardwoods, Mr. P. Palmer, C.E., says :—"The properties of the Australian Karri and Jarrah woods are such that they are almost perfect as materials for paving purposes ; they are extremely dense, and only absorb a small amount of moisture, and when the blocks are laid on a bed of concrete, and the blocks thumb-dipped with the same material, and laid with a cramped, joint—that is, one as close as the blocks can be cramped together, a roadway almost impervious to the absorption of surface impurities can be made."

The following is quoted from a recent issue of the *Illustrated Carpenter and Builder* on the sanitary advantage of wood pave ments for streets :—"With respect to the sanitary conditions of wood-paved streets, there ought to be no substantial reason against classifying deal and soft wood pavings as noxious, and hardwood generally as innocuous, the Jarrahdale taking, for sanitary conditions, foremost rank. The wood itself is so impervious that heavy rain rapidly runs off from the surface, which is dry again in a few minutes after cessation of the shower. It is also so thoroughly non-absorbent that the noxious and offensive fluids which are generated by the solutions and decomposition of organic refuse drain harmless away. In a lecture at the Parkes Museum of the Sanitary Institute, Mr. Charles Mason, the Surveyor of St. Martin's-in-the-Fields, dealt with one serious aspect of the subject, the scavenging of the different pavements. Asphalt he puts first for facility of being cleansed, but liable in certain conditions of weather to become greasy and slippery. Granite is very generally being supplanted by wood on account of the noise produced on the granite by the traffic. It is, however, where properly laid and maintained, well adapted for efficient scavenging. 'Wood pavements,' he said, 'had been the subject of serious complaints on account of their absorption; and there is no doubt the opponents of this class of pavement have a strong case against wood when it is allowed to remain in a bad state of repair, and consequently cannot be properly scavenged. The system of laying the blocks close together ensures a more even surface, consequently one better adapted for cleansing. The hardwoods now being introduced into this country afford a very good example of a pavement upon which a fairly good foothold for horses can always be secured, and when the blocks are evenly and truly cut, and laid close joined, a practically impervious pavement is obtained without the risk to horses that is so prejudicial to the use of asphalt.' "

CHAPTER XVI.

ASPHALT AND TAR MACADAM.

NATURAL asphalt is a form of limestone or sandstone in union with mineral pitch or bitumen. Bituminous limestone is a rock naturally combined with bitumen to the extent of from 5 to 15 per cent., while bituminous sandstone contains bitumen in varying quantities up to 50 or more per cent. These deposits are to be found in considerable quantities in some localities. It is obtained principally from Val de Travers, Switzerland; Lemmer, near Hanover; Brunswick; Ragusa, Sicily; France, Germany, and Trinidad.

The Val de Travers asphalt is brought from the mines into this country in its raw state, where it is prepared for use in the manner requisite for different purposes. For street pavements it is ground to a very fine powder, and after being heated, is conveyed in a semi-plastic condition on to the works in covered carts. The material is then evenly spread by means of rakes over the surface of the bed provided, and compressed with heated rammer, or "pilons," and afterwards smoothed off with smoothing irons or rollers. The finished thickness is generally 2in. The foundation upon which this asphalt is laid consists of 6in. to 8in. of concrete, similar to that adopted for wood pavements.

Mastic asphalt.—In this method the rock, as imported, is crushed into small pieces, and to this material is added a small percentage of pure bitumen. The mixture is then heated in cauldrons on the site of the work, and when sufficiently melted, dry clean sand and gravel are added, and thoroughly boiled in the liquid for an hour or two. The material is then run out over the surface of the foundation previously prepared, and smoothed down with hot irons or rollers.

Trinidad Asphalt.—This asphalt exists in three descriptions of deposits. Lake pitch is from a deposit covering some acres of land, and of great depth; in fact, liquid material continues to be thrown up at one point, which is said to be the mouth of an extinct volcano. Land pitch is that material or deposit resembling lake pitch, which is found distributed over the surrounding land, and generally covered over with formations of earth of varying thicknesses. Iron pitch consists of certain portions of the lake and land pitch, which have been rendered very hard by some unknown cause, supposed by some authorities to be the result of heat on the softer pitch, caused by forest fires. For street pavements a compound is made of Trinidad pitch, broken stone, sand, and other materials, which is ground and laid in a heated powder, similar to that already described.

Barnett's Asphalt is an asphalt of which the resisting materials consist of small particles of iron ore. This is laid in a similar manner to that of mastic asphalt.

There are other systems of asphalt pavements that might be described from exhaustive trials that have been carried out from time to time in London and elsewhere; however, the foregoing are all that are required for general purposes—the preference being given to Val de Travers.

Tar Macadam.—The usual method adopted in preparing tar macadam for roads is to heat clean macadam stones on an iron plate with a fire under, and whilst in a heated state to place them on a mixing surface in circular heaps of about half a yard cube; the centre of the pile is then hollowed, and into this is poured a boiling mixture of tar, pitch, and creosote oil, which has been boiled together in a cauldron in the proportion of one barrel of tar, 1 cwt. of pitch, and four gallons of creosote oil. The stones and tar are then well mixed together by being turned several times. This mixture upon cooling is ready for use, and is laid on the dry road bed and rolled in three layers, the two first layers being 3in. thick, and the topping layer 1in. thick, of smaller materials.

The author's experience is that coal tar well boiled in a cauldron equally answers the purpose for road-making, as compared with the addition of creosote oil and pitch. It is, further, unnecessary to heat the stones, unless they are wet, in which case they should be thoroughly dried, but not overheated. The author found that the life of some granite tar macadam laid by him to be considerably reduced by the heat of the hot plates in drying.

Tar macadam roads require to be kept well sanded for a few days after the traffic is admitted over them. Tar macadam is an inexpensive substitute for channelling at the sides of country and lightly-used roads. In tar macadam roads considerable care is necessary on the part of those responsible that the road bed upon which the material is placed is perfectly dry, otherwise the work may be a failure. Streets laid with tar macadam, when the work is properly executed, are sanitary, noiseless, and inexpensive roads. They are, further, easily kept clean, and afford general comfort to horses.

CHAPTER XVII.

BRICK PAVEMENTS FOR ROADS.

THE essential points to be considered in selecting a paving brick for roads are that they should be constructed from a good fine clay practically free from lime, and one that will stand high burning, in order that the brick may become sufficiently annealed to render it hard, tough, and impervious. The bricks should be as uniform in quality as possible. They should have a resistance to crushing of at least 7500 lb. to the square inch, and should not absorb more than 2½ per cent. of water after twelve hours' immersion.

A brick that appears to be impervious on the outside is not always so on the inside, so it is always desirable to break the brick before immersing in water. When one brick is thrown on to another, should the edges chip the brick lacks toughness. Porous or soft bricks should never be used for paving purposes, as they are very liable to become insanitary, to wear away very quickly, and to become uneven. Good bricks, well laid, make a fairly durable, sanitary, noiseless, and safe pavement.

The City Engineer of Chicago, in comparing brick with asphalt, says :—" I believe that for general use shale brick, properly burned and of the right size properly laid on a hydraulic cement concrete foundation, as superior to asphalt, as to its first cost, facilities and cheapness for repairs, its sanitary qualities, its ease upon horses, and last, but not least, its durability. Brick can be laid on a grade where it is out of the question to lay asphalt. . . . Brick when wet is not more slippery than when dry; asphalt is always dangerous, and if wet is more slippery than when dry, and when a horse is down on asphalt it is with difficulty that he regains his feet. Brick is not more noisy than asphalt, and as it can be continually sprinkled it is far less dusty."

Brick pavements require an unyielding foundation, so that the small blocks shall not be forced downwards by the weight of the traffic. A foundation of concrete similar to that recommended for wood and asphalt pavement is the most suitable. The thickness of the concrete should depend much on the nature of the traffic and the street where laid. Where the traffic is light and the road bed sound, a thickness of 6in. should be sufficient, but should these conditions be departed from, then from 8in. to 10in. may be necessary.

Brick pavements are laid on a thin layer of mortar, or sometimes well-consolidated sand spread over the concrete foundation. The bricks are laid on edge as close together as possible. Commencing at the kerb, stretches are laid crosswise with the street in a straight course, to meet the kerb on the other side of the road; the second course is commenced with a half brick instead of a stretcher, and the remainder carried on as before with whole bricks. The transverse joints between the bricks by this means become broken, and in like manner the pavement should be laid throughout. At the completion of every few yards of paving the surface should be carefully rammed till the bricks are level. Any brick that may have been compressed too low thereby must be removed, and more mortar or sand added, and the brick replaced. On this ramming being done, the joints should be grouted in with cement, liquid cement being spread over with a broom till the joints become quite filled. Previous to the traffic being turned on to the pavement, a coating of sand should be spread over the surface.

CHAPTER XVIII.

STONE PAVEMENTS.

STONE pavements of granite sets are unsurpassed for durability and cleanliness, but from almost all other points of view they are objectionable. For the purpose of cab stands these pavements take the lead, and are beaten by no other material for this work.

Granite setts, cubes, or blocks, wherever employed, after a time wear very smooth and become very slippery; they are then not only a danger to traffic, but also a great and cruel strain upon the horses, especially at times of starting with heavy loads. The pavement further constitutes such an immensely hard and unyielding construction that horses not only receive injury from the strain in their efforts to maintain a foothold, but their feet and legs give way through the severe shaking they receive each time their shoes come in contact with the rigid surface. If such surfaces are well roughed then they are rendered almost unbearably noisy, besides being unpleasant for all light traffic using the road. Granite setts for paving purposes should not exceed 4in. across, with a depth of from 7in. to 9in. The lengths should be equal, but not greater than 12in., and they should be laid so as to break joint with their lengths across the road. The setts should be placed on a foundation of 6in. of cement concrete, and bedded in an inch thickness of fine gravel or gas tar asphalt, and on completion should be thoroughly rammed to a level surface. The joints should be filled in with fine gravel, and afterwards thoroughly grouted with a mixture of hot tar and creosote oil until quite full. On completion, before the traffic is admitted on to the new work, a coating of dry sand should be spread over the surface.

Cement grouting in the place of pitch is sometimes employed, but the advantage in the use of the latter is that the tar gives a certain amount of elasticity to the pavement, which is most desirable in reducing the liability to fracture, which would render the pavement pervious to water. At all junctions of streets the setts should be laid diagonally (Fig. 33, Chap. XIV.), so that the traffic, in crossing from one street to another, should not follow the unbroken joints, which would sooner or later form ruts; and, at the same time, this method affords a better foothold to horses turning corners of streets than would otherwise be the case.

Sandstones.—Good close-grained sandstones make a less slippery and noisy pavement than granite, but are not so durable a material. The wearing qualities of sandstones vary considerably. Some formations, such as those which are dense, homogeneous, and tough, give very satisfactory results, whilst others which are more permeable to water have little wearing properties, which are further reduced upon their being exposed to the weather and frost.

Cobble Stones.—Street pavements of cobble stones are now almost a thing of the past, although in by-gone days such pavements were much employed. These pavements consist of rounded pebbles, from 3in. to 5in. in diameter, which are set on end, and bedded in a layer of concrete. The upper portion is then filled in with fine gravel. More often than not cobble stones are set in position on a bed of gravel or sand, and when laid the stones are well rammed, and on completion the upper interstices are filled in with a layer of fine gravel or sand, to be afterwards worked in by the traffic. This is an inexpensive method of paving, but does not constitute a suitable surface for many of the modern modes of locomotion. The only other advantage it has besides cheapness is that it affords a good foothold for horses.

CHAPTER XIX.

OTHER IMPORTANT PAVEMENTS.

The Bingham Patent Paving (Fig. 35)—This pavement was invented by Colonel John Bingham in 1895. After a serious accident that befel this gentleman whilst driving through one of the stone paved streets of Sheffield, he was led to investigate the question of the use of stone for street paving purposes. His conclusion was that granite was highly dangerous and unsuitable for paving purposes, and if employed it should be combined with wood. He afterwards designed the paving known as the Bingham paving, which has since been used with much success in the principal streets of Sheffield, and from the interest that is being taken in this combination it is more than probable that it will become one of the leading pavements of the future. The advantages claimed for this pavement over many other materials, and methods of laying, may be quoted from an interesting article contained in a recent issue of the *Sanitary Record and Journal of Sanitary and Municipal Engineering*:—

"The method of laying is comparatively simple. First, a good concrete foundation must be secured, after which the laying of the blocks is proceeded with in the same way as is generally adopted with granite setts or wood. The blocks should be laid close and the wood strips be dipped in pitch. The section on London-road gradient was grouted with pitch, but it has not proved satisfactory, so that it is considered best by experts that the grouting used should be cement and sand. The wood is laid at the sides and ends of the granite setts. The depth of the wood blocks varies with the granite—that is, if new 6in. setts are used the wood must be fully 6¼in. deep; but if old setts are used the depth of the wood may vary accordingly. The length of the wood blocks is 9in. for the sides of the granite setts and 3½in. for the ends, with a width of 1½in. These sizes have been found to be generally the best for

Fig 35.

the purpose. One distinct advantage is that the class of granite used is immaterial, as the objection to the harder sorts of granite pitching is removed in the combined pavement. Granite which has been used, being smoother than new granite, is in some respects preferable, as it can be more easily swept clean of manure, &c., which is all the cleansing required. As already stated, the most suitable wood is Tasmanian stringy bark, which, in its durability and well-known antiseptic character, compares favourably with any of the Australian hardwoods. The especial advantage of stringy bark arises from its toughness, which renders the blocks less slippery than some of the hardwood blocks and causes the wood to " burr." By the use of wood in combination with stone, the difficulty arising from the contraction and expansion of the wood, which sometimes occasions so much trouble in wood-paved streets, is avoided altogether.

Among other advantages claimed for this system are :—

(1) Safety, inasmuch that, while possessing the wearing quality of granite, it is non-slippery, and is therefore safe for every kind of street traffic. This fact, in view of the frequent and serious accidents which occur in granite-paved streets, is one of considerable importance.

(2) Durability, as already proved by the heavy tests in Sheffield.

(3) Cheapness, as it is cheaper than wood, and ultimately far cheaper than granite. It is difficult to estimate the cost of the pavement, in view of the varying prices of material and labour in different localities ; but when adequate supplies of stringy bark are obtainable, the cost of foundation, labour, and material should not be more than 14s. to 15s. per yard super for 6in. setts, 13s. to 14s. for 5in. setts, and 12s. to 13s. for 4in. cubes, with the necessary wood intersections. The prime costs of a new all-granite pavement on the present advanced prices would be from 13s. to 14s. per yard super of 6in. setts. Granite setts, however, after a few years' wear, need re-dressing—an expense that is avoided in the Bingham pavement, as the wear is much more even, and the pavement being non-slippery, no re-dressing of the granite is required. Perhaps the most important saving is effected by utilising the old granite setts, which would otherwise have to be broken up, but which, for this purpose, are as good as, if not better than, new granite. In the event therefore of re-laying an old granite-paved

street with the Bingham pavement, the cost, including new foundation, should not be more than from 8s. to 9s. per yard super. The cost of maintenance of this class of paving is comparatively low. In the first place, very little supervision is required, and, secondly, no grit or sand is needed, and consequently little or no watering. Granite-paved streets, to be safe, required to be constantly shingled, at a cost of between £110 and £120 per mile of street per annum; while the dust created by shingling has to be laid, in dry weather, by constant watering, which involves a cost of about £20 per annum—or a total cost of 1s. 6½d. per yard run per annum. The saving of expense in these particulars will be considerable, as well as in the cost of clearing drains, choked by dirt and street detritus, which, accumulating on the surface, under the ordinary conditions of granite streets, are washed into the gutters and drains. The patent paving costs no more for sweeping than asphalt. The non-use of sand or grit minimises dust in fine weather, and dirt in wet, and although less scavenging is required greater cleanliness is secured.

W. Duffy's Patent System of Dowelled Wood Paving.—In Chapter XIII. it was stated that the wood that absorbs the least quantity of moisture is the one to be preferred from a sanitary point of view. The use, however, of the harder and more durable material renders it more necessary than ever in obtaining a perfect pavement to have a system of effectually bonding the blocks together, so as to prevent the possibility of their becoming loose, tilting, or rising. This dislocation, and consequent tilting and rocking of the blocks, when allowed to continue, will have a deleterious effect on the life and cleanliness of the pavement. A means of ensuring the attainment of these essential desiderata is presented under Duffy's patent system of dowelled wood paving. It is claimed by the inventor that under his system the blocks are bound together, so that a homogeneous and practically immovable pavement is obtained, even when subjected to the hardest vehicular traffic. A road pavement laid under this system is extremely simple, and is inexpensive to apply, especially as shallower blocks than usual may be employed with good results. It is also claimed for this system that by its use it may even be practicable to dispense with the concrete foundation, which need only be used in streets subjected to heavy traffic.

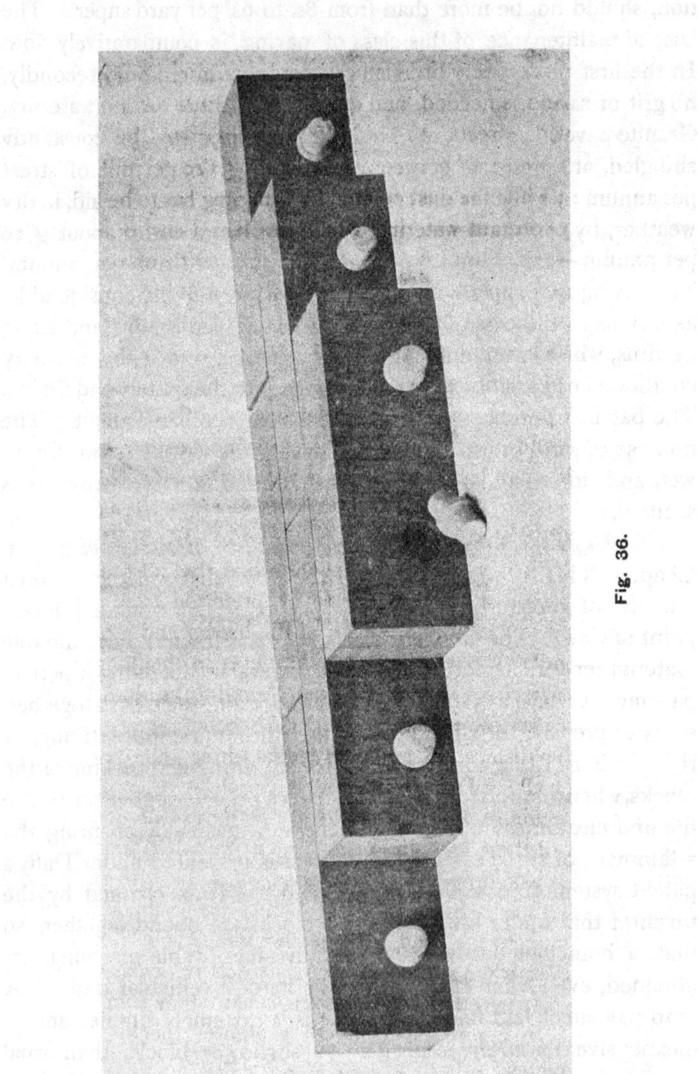

Fig. 36.

The method employed by Mr. Duffy in securing the blocks together consists essentially in the forming of countersinks, or recesses in the blocks, into which are inserted dowels, or bonding keys of a special type—as shown in Fig. 36. It appears that this system of dowelled wood paving has long since passed the experimental stage. It was adopted by Sir J. Wolfe Barry, K.C.B., for the roadways of the Tower Bridge, which are in a perfect state of preservation, after over six years' exceptionally hard wear. It has also been laid with good results at many important railway approaches and streets in London and the provinces.

McDougall's Patent "Combination Paving Setts."—This combination paving sett consists of a thoroughly vitrified paving block (10in. by 4½in. by 4½in.), which is of a very impervious

Fig. 37.

nature. It is pressed and stamped under great pressure, rendering the block immensely strong. Into each block are inserted ten wooden plugs 2in. long and 1in. square, which extend only partly through the sett, and are wedged in so that they should not work loose. Contraction and expansion of the wood are provided for. These blocks, although sufficiently strong in themselves to carry heavy traffic, should be based on a layer of concrete 4in. in thickness to keep them firm. The advantage claimed for these setts is that a good grip is afforded by the wood plug, which projects from the face of the block about $\frac{1}{16}$in. The wood plug consists of thoroughly creosoted Baltic spruce of a long, tough fibre, set on end. It has been proved after repeated trials that the soft wood plugs are much more satisfactory and durable than hard wood for this purpose. The fact that the soft wood slightly gives and spreads under the weight of the traffic renders it much more desirable for the purpose intended than hard wood, which is too rigid when subjected to compression.

The author has taken part in substituting these setts in the place of granite crossings, and this resulted in a decided improvement.

Mr. J. Hall, C.E., has laid many such crossings at Cheltenham in the place of previous granite ones, with much good effect. Fig. 37 is from a photograph of a block taken up from a busy street after six years' wear.

Concrete Macadam.—This is composed of broken stones with an admixture of cement and sand. These materials are mixed together in the proportions of 1 of Portland cement to 1½ of sand and 4 of clean broken stone, and after being well mixed together are placed on the road bed, which must be previously formed and consolidated by rolling. The thickness required for this material greatly depends on the description of traffic for which the road is intended. For ordinary traffic a thickness of 7in. in the middle and 5in. at the sides is sufficient, but for heavy traffic a thickness of 8in. is recommended. The concrete should be placed on the road in two layers not exceeding 4in. in thickness, and each layer properly rolled by a hand roller weighing about 1 cwt. These roads are very durable and easily constructed.

Sanitary Block Pavement.—The sanitary block pavement, which has recently been introduced into this country from the United States by the Hastings Pavement Company, consists of asphalt blocks made of a mixture of refined Lake pitch and crushed trap-rock, which are incorporated whilst at a temperature of 200 deg. Fah. At this temperature the material is manufactured into blocks under a pressure of 120 tons each block. The dimensions of these blocks are 4in. by 12in. by 3in., and 4in. by 12in. by 4in., and they weigh 13½ lb. and 18 lb. respectively. This pavement provides a long-felt want of a sanitary and non-slippery material for streets, and no doubt has a good future before it in this country. The quarries from which the trap-rock is obtained for the manufacture of these blocks have been inspected by the author, and he is well acquainted with the tough but non-slippery nature of the material when used as road macadam, but apart from that, on the block pavement, being laid in 4in. courses, the joints themselves are sufficient to check any tendency of horses' shoes slipping on its surface. The pavement is further considered to be perfectly sanitary, the materials when properly laid constituting a roadway impervious to moisture or noxious matters. The sanitary blocks are laid, as illustrated in Fig. 38, on a concrete foundation in practically the same manner as that described for wood block pavements, only no expansion joints are necessary.

Roads with Stone Trackways.—Stone trackways are sometimes employed in streets where the traffic is heavy, for the purpose of minimising the difficulty of traction—constituting a firm and solid

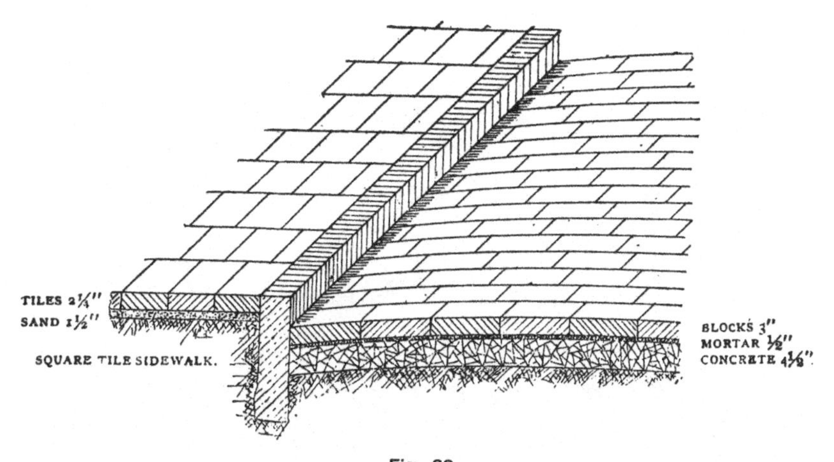

Fig. 38.

wheel track which, in narrow roads, where all the traffic has to follow one line, is of great value in preventing the road being cut into ruts. The space between the wheel tracks may with advantage be constructed of rougher materials to afford a good foothold for horses. Trackways should be constructed of smooth blocks of granite or sandstone from 18in. to 24in. wide and 4ft. to 7ft. long and 6in. to 9in. deep. They should be laid perfectly flat, with their ends close together. The blocks are usually laid on a coating of fine gravel spread 2in. thick over the ordinary road bed, but concrete is to be preferred in places where the traffic is heavy. The interval between the tracks should be 2ft. 4in. in width, constructed of cobble stone, rough tar macadam, Bingham's paving, McDougall's combination bricks, or some other non-slipping material. Such tracks placed on hills, with a non-slipping footway, diminish the amount of traction force, and render a steep incline more easily ascended with heavy loads.

CHAPTER XX.

BRIDGES.

Details for a Girder Bridge.—Road bridges for general traffic are liable to a great variety of strains. Provision has to be made in country districts for carrying heavy machines, such as steam ploughs, traction engines, and in towns, steam rollers. In these cases the loads are necessarily concentrated, and call for the closest attention in designing the details of a bridge.

The main girders of a bridge support the flooring, and any point in this flooring—in a bridge for general traffic—may become the point of application of any concentrated load. It is, therefore, imperative that each subsidiary girder which forms part of the floor shall be designed to support such concentrated load on any part of its length, and that the flooring between these girders shall be strong enough to support it. A bridge, to be properly designed, must have the stresses calculated, beginning at the roadway and passing through all the subsidiary girders to the main girders which carry the whole work. The floor may consist of small transverse or cross girders, supported at each end by the main girders on either the top or bottom flanges. If the main girders are to be made of solid I section, the cross girders will be put about 4ft. apart, and the intermediate space filled up with buckled or other stiffened plates. When the main girders have triangular web (of lattice pattern), it is necessary to connect the cross girders with the flanges at point of junction of the struts and ties with the flanges, so that no transverse stress shall be brought upon the flanges. In large structures the distances between these points will be too great for filling in with plates, and, therefore, intermediate longitudinal girders will be required between the cross girders, and the filling in plates will be carried by the intermediate girders.

There is another kind of longitudinal girder used when the

cross girders are not far apart, and its duty is to distribute concentrated loads over several cross girders. For instance, a bridge which has its cross girders 4ft. apart, and subject to the passage of a heavy vehicle which has its wheels 8ft. apart longitudinally. When the front and back wheels are on two cross girders there will be one in between which receives no load at all, and those beyond the wheels will also be unloaded. The two loaded cross girders will deflect under the loads which come upon them, and if all the cross girders are connected by a light longitudinal girder, riveted above or beneath them, the otherwise idle girders will receive a part of the load for the intermediate girder, which must deflect with those on each side of it. The longitudinal girders are termed "distributing girders," and do not carry any part of the load to the piers.

The forms of bridge floors most generally in use may be better understood by reference to the accompanying details and illustrations. Design No. 39 shows a road bridge constructed by Messrs. David Rowell and Co., with rolled steel girders of I section and buckled plates. The span of the bridge is 20ft. and the width 10ft., and made to carry a weight of from 15 to 25 tons. The main girders, which are four in number, are of 12in. by 6in., by 54 lb. per foot, and are connected by $\frac{5}{16}$in. buckled plates, having a camber of $2\frac{1}{2}$in., and a flange formed on both sides of same, $2\frac{1}{4}$in. deep, for bolting to web of girders. These plates are in lengths of from 3ft. to 4ft., as desired, butted together. (It has been found that, with a buckled plate riveted down to its bearings, all round its edges is twice as strong as the same plate merely resting on its edges.) The joint is often made when the two plates merely butt together by using a cover plate or strap of T section iron, making what is technically known as a single butt strap joint; this then strengthens the plates at their weakest part. It is advisable in a bridge constructed of R.S. girders to connect the two outside ones together by means of ties. In this design there are three in number, of 1in. diameter, screwed at each end, and tightened to suit width of bridge.

Timber flooring without any ballast is often used for foot-bridges, and sometimes for private road bridges, when it is desired to keep the first cost as low as possible, though it cannot be recommended for first-class structures. The hand railing is usually carried in this class of bridge by a steel tee-iron or cross-section cast iron standard, bolted to the top flange of the outside girder, as shown in Fig. 39. In this case it was made of $1\frac{3}{4}$in. diameter piping running through holes drilled in the 4in. by 3in. by $\frac{1}{2}$in. tee standards.

Design No. 40 is an example of steel troughing construction, which is of comparatively recent origin, and is now used extensively for bridges. The features of merit possessed by this system are :—

(1) It produces an even distribution of the load on the main girders instead of a series of concentrated loads at the points of

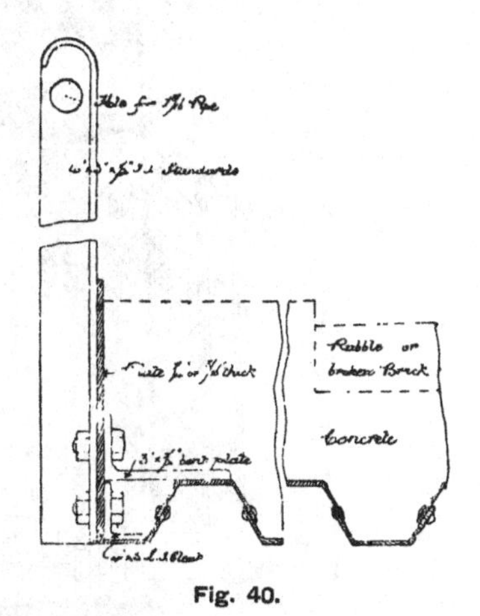

Fig. 40.

application at the cross girders. (2) It dispenses with the necessity of cross girders and bearers, and its small depth gives increased head room, which in many cases is an important consideration.

The troughs are generally constructed of mild steel capable of sustaining an ultimate tensile stress of not less than 30 tons per square inch, with an elongation of 20 per cent. The troughing rests on and is riveted to the main girder, or to an angle or tee iron riveted to the web of the girder, or for bridges of small span, where it has a bearing of 9in. to 12in. on either side of the span. The main girder can be done away with, and the flooring can be bolted to a level concrete or stone bed by lewis bolts at about

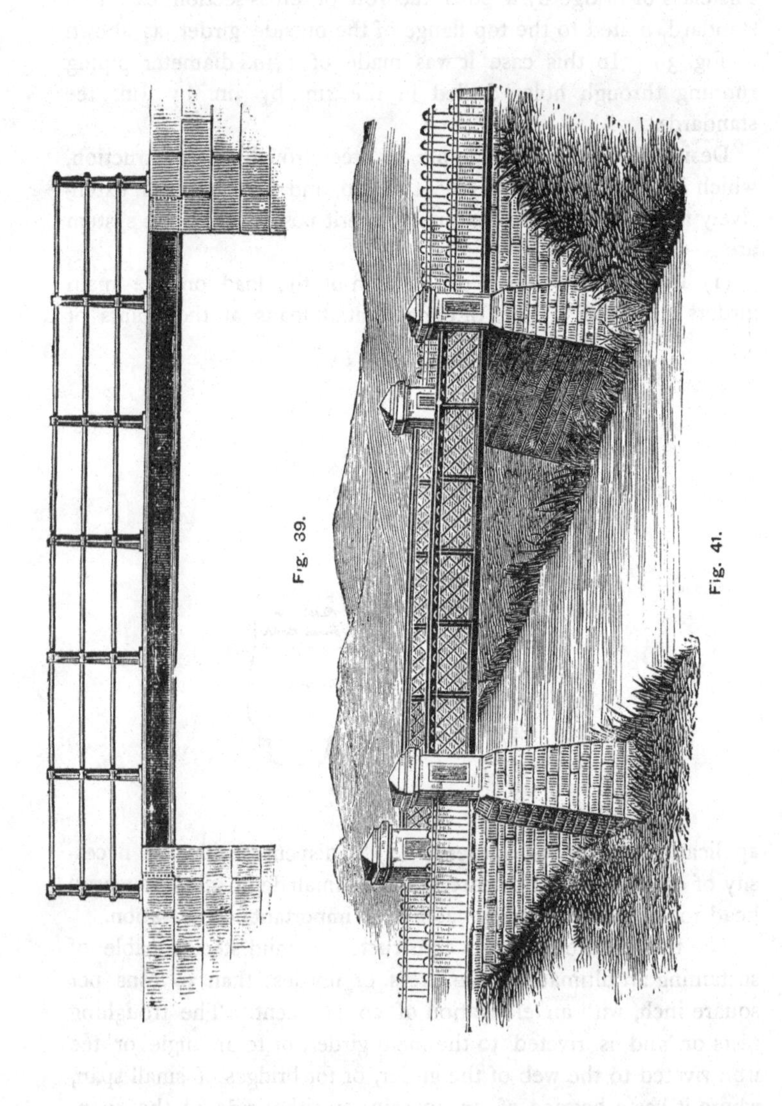

Fig. 39.

Fig. 41.

from 3ft. to 4ft. centre. The hand railing in this instance is fixed as in last design, only it is secured to the bottom of the troughing by an angle cleat and a flat bar on top.

The flooring of a bridge of design No. 41 of the same span and width as last, was constructed of steel troughing of section, weighing 35 lb per foot super of floor space covered, which had a bearing of 12in. on each end, and was bolted on to a flat bed of

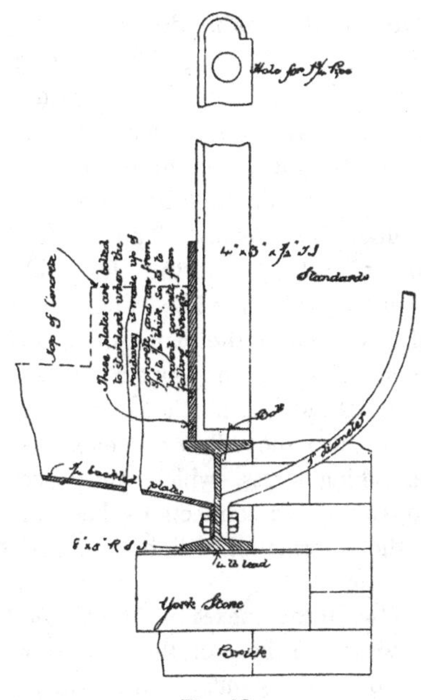

Fig. 42.

cement concrete on a pad of 4 lb. lead, and was kept in place by ⅞in. lewis bolts at 4ft. centres (Fig. 42).

The roadway of a bridge is usually filled in with concrete about 4in. to 6in. above the top of the girder or troughing, as the case may be; the whole can then be covered with stone or other suitable material.

Lattice Girder Bridge of Design No. 43.—When a lattice girder of a given span and depth is loaded uniformly the stresses on the braces are universally proportional to the numbers of systems of triangles in the web. For example, the stresses in the braces of a girder like that shown in design No. 43 are approximately one half of those in the corresponding braces in a girder of a single system, while in a girder of a quadruple system like that shown in design No. 41, they are approximately one-fourth of those in No. 43. Lattice girders usually have vertical members situated at certain intervals along the girders; these are introduced with the object of distributing the load between the top and bottom flanges, and also of giving lateral stiffness to the girder. As an example, take a triangular web girder bridge, as in design No. 43, which has a clear span of 85ft., and an effective girder span of 90ft. The web is made up by bars, which, with the flanges, form a series of equilateral triangles. The girders are each divided into thirteen triangles. The load is brought on to the bottom flanges at the points of connection with the web bars, so that no transverse stress shall come upon the flanges. The bridge is required to carry a single line of railway, and the clear width between the flanges of main girders will be 16ft. The connections between the flanges and the web bars of the main girders will be made by pins, which are more suitable than groups of rivets for the purpose, because they do not cut away so much sectional area in proportion to their strength, and also they allow of deflection in the girders without incurring raking stress, which must occur when joint plates are used to make these connections, because in the ordinary course of matters the deformation will be attended by changes of the angles of the joints.

The distances between the apexes of the triangle formed by the centre lines of the webs and those of the flanges will be 6ft. 2in., and that will also be the distance from centre to centre of the cross girders. This distance is too much to be spanned by floor plates or longitudinal sleepers, and therefore these rail-bearing girders between the cross girders and the flooring between them can be made up with floor plates, or the shallow trough flooring described in design No. 40. The floor girders are made with a depth of about one-twelfth of the span. The floor plates at the side are supported upon angle iron riveted to the bottom plates of the main girders. The portion of the floor carried upon one rail bearer is half that between it and the main girder, plus half that between it and the other rail bearer.

In girders of small depth, a good form of flange consists of

Fig. 43.

Fig. 44.

horizontal plates with deep angle irons between which the ends of the struts and ties are held. The struts in such cases may conveniently be made of tee iron or steel riveted together back to back with a plate between them if necessary, and for very light girders angle iron or steel may be used placed back to back, where they cross for both struts and ties. In those cases where the series of triangulation are so numerous as to form a lattice web, the flanges can be calculated in the same way as those of a plate girder. When the ties and struts cross each other it is usual to rivet or bolt them together at their intersections, which helps to give lateral steadiness; but this must not be taken as shortening the effective length of the struts in regard to their capacity for resisting compressive stress.

In cases concerning bridges of longer span, the connections between the web members and the flanges will often be made more conveniently by means of rivets than by the large pins which would become necessary, as these would necessitate very heavy thickening plates for bearings, which rivets do not require. The great objection to these joint plates in small girders is that they disfigure them, but in the larger structures they are not so conspicuous. No general rule can be laid down, but the designer must decide according to the circumstances of each case what forms of struts and connection will be most suitable for his purpose.

Details of Suspension Bridges.—The commonly recognised form of suspension bridges for foot passengers is shown in accompanying illustration—Fig. 44—as manufactured by Messrs. D. Rowell and Co. The suspension and roadway cables are of strong patent steel galvanised wire, which is continuous, running over saddles of cast iron attached to top of timber columns—or, if preferred, to ornamental columns constructed of lattice steel construction—and anchored at each end behind a mass of buried masonry. Vertical suspension rods carry the floor of the bridge. These are of solid steel, and secured at each end to the cables by means of special iron clips. The stresses upon the suspension cable amount to the same as those of an arch of the same proportions; but they are tensible instead of compressive. Of course, the cable may be run over a number of consecutive spans if required before being anchored.

In an ordinary bridge with vertical suspension rods, when a heavy load passes over it, a slight deformation of the structure sideways takes place; this can be prevented to a certain extent by designing the bridge with the suspension rods slightly inclined in an inward direction to the vertical, when an element of stability is introduced, because there will be initial tension on the suspension rods in addition to that due to the direct floor load, and oscillation occurs, displacing the suspension rods. The initial stress will be diminished on the rods of one side of the bridge, and increased in the rods on the other side; this, then, will tend to bring the floor to its normal position.

Bridges of this design are specially suitable for spanning cuttings and streams near towns. While being pleasing in appearance and simple in erection, their cost is also considerably less than girder bridges.

CHAPTER XXI.

FOOTPATHS.—NATURAL AND ARTIFICIAL STONE FLAGGING.

FOOTPATHS, like roads, require a good deal of attention to ensure durability, safety, and comfort. Footpaths differ from carriage-ways in the essential that whilst one is constructed to carry the heavy weights of vehicular traffic, the other is intended for the lighter use of pedestrians. The essential requirements of a good footpath are that the foundation should be firm and unyielding, the paving material smooth and tough but not slippery, of fine texture and uniform quality throughout. The material should, further, absorb as little moisture as possible, should not be of a slaty nature so as to flake, and should not cause a glare at times of sunshine that would be hurtful to the eyes. The form of the path should be such that will give a uniform slope of from $\frac{1}{4}$in. to $\frac{1}{2}$in. in every foot towards the kerb, so that the water will freely run off its surface, whilst its width should be sufficient to comfortably accommodate the general requirements of the district. The model by-laws of the Local Government Board prescribe that the person laying out a new street "shall construct on each side of such street a footway of a width not less than one-sixth of the entire width of such street." The materials used for footpaths are natural stones, consisting of granite slabs, Yorkshire and Caithness flagging, and flagging of limestones and sandstones. Also artificial stones, consisting of Victoria, Adamant, Non-slip, Imperial, Elliott's, Ransome's, Bucknell's, Stuart's granolithic, and Jones's annealer, which are slabs for paving purposes made of the various materials employed by the manufacturers of the foregoing stones, and moulded, pressed, or otherwise treated according to the particular method adopted in each case. There are in addition concrete paving laid *in situ*, mastic asphalt paving, tar paving, brick paving, and footpaths made of gravel.

Granite Slabs.—Granite slabs of various thicknesses are much employed in granite districts, and sometimes in other parts of the country where the paths are subjected to exceptionally heavy wear and tear. Granite possesses many points in its favour as a paving material for footpaths; it is exceptionally durable, which often more than makes up for its first cost, which is usually greater than that of other paving materials. Granite, where the traffic is heavy, requires roughing occasionally, or it wears slippery; but when roughed at intervals of from five to seven years, according to traffic, it affords a good foothold. The cost of roughing at Penzance is 1s. a yard super. As a rule, no separate kerb is employed to paths where granite slabs are used; the outer slab is so dressed that one edge forms the kerb. When granite slabs are worn too far to be used any longer as a footpath the material is always useful, and will still be good value for many useful purposes, and at the last can be broken up for road metal.

Yorkshire Flagging.—This flagging perhaps stood by itself amongst paving materials for excellence for many years; but since the introduction of the Victoria and other artificial stones, its use is not so indispensable to the construction of a good flagged footpath. York stone combines durability and firm foothold, is pleasant to walk upon, and can be easily cut to any required size. Its disadvantages are that the lamination causes it to be liable to flake, especially in frosty weather; the material, even when specially selected, often wears unevenly, so that a soft flag wearing away more quickly than a harder one causes a pool of water. In order that York may be turned after its one side is worn hollow, both sides need to be dressed. York stone flagging should be 2in., 2½in., or 3in. thick, and no flag should contain less than from 6 to 8 superficial feet. York stone of the "Rivingstone" class, with a greyish colour, should always be used The flagging should be set flush on 3in. of clean grit or sand, and the joints made with cement grout or mortar.

Caithness Flagging is a stone obtained from Thurso, N.B. This stone is equal to, and in some respects superior to, York. It possesses most of the desirable features of a good paving material. It is most durable, its wear is even, and it does not flake, and is not affected by frost. It will stand rough usage—even light vehicular traffic will not injure it; it does not wear slippery, it absorbs little moisture, and gives excellent resistance to bending stress. The natural faces of the stone can be used, which saves cost in labour, and the thickness of the flagging need not exceed from 1½in. to 2in.

Sandstone from Fishponds, near Bristol, is used with good results as a paving material. Purbeck, or limestone, obtained from Dorsetshire, is another excellent material, but it is inclined to wear a little slippery. Devonian limestone is also much used for paving footpaths, and gives good wearing results.

Patent Victoria Stone.—The introduction of this artificial stone in 1868 afforded a material that would as nearly as possible meet the requirements of a perfect footpath, combining as it does the important features of being most durable, impervious, non-slippery; of a neat and pleasing appearance, easily laid, and as good on one side as on the other. Since that date the company working the patent state that over 500 miles of footways have been paved with the Victoria stone, and at the present time their output is from 30,000ft. to 40,000ft. of material per week, and in order that all stone should be thoroughly seasoned before leaving the works, as much as three to four million feet is always kept in stock.

The composition of this stone, in the words of the manufacturers, is "an amalgamation of granite and Portland cement, steeped by a patent process in a solution of flint." Mr. H. Reid, in "A Practical Treatise on Natural and Artificial Concrete," gives the following description of the process by which patent Victoria stone is manufactured:—

"Three parts of aggregate are thoroughly mixed in a dry state by machinery, and the water then added in a careful manner, so as to avoid the danger of washing out any of the fine and more soluble portions of the cement, and before any initial set of crude concrete can arise it is put into the moulds, in which it is worked with a trowel so as to fill up the angles and sides, thus ensuring accurate arrises all round. . . . . Slabs are made of various sizes, but it is found that the most useful sizes for London paving work are 2ft. 6in. in length by 2ft. wide, and 2ft. square by 2in. thick. Paving slabs of this size weigh 25 lb. to 26 lb. the foot super, and are convenient to handle. The moulds, filled in the manner thus described, are allowed to remain on the benches of the moulding sheds until the concrete has sufficiently set and so much of the water of plasticity evaporated as will permit the slabs to receive the beneficial influences of the silicating operation. This indurating process is one of absorption, and the best practice is that which provides a reasonably porous mass, to which may be introduced an accurately prepared liquid silicate of the desired specific gravity. The slabs, when sufficiently dry, are relieved from the surroundings of the moulds, which, being made in pieces, can be readily detached by unscrewing the fastenings. The slabs are then taken to the tanks in the silicating yard (protected from the weather), placed side by side, and covered by the silicate solution, where they remain until the proper beneficial influence has been duly imparted. The period of time required to complete the silicating process is not of a fixed or arbitrary character, and depends on the condition of the slab and its capacity of absorption. About fourteen days, under ordinary circumstances, is regarded as sufficient to secure the desired advantages of the process."

The wearing qualities of the patent Victoria stone are excellent. In London it has been found to be more durable than York. According to the *Metropolitan*, some Victoria and York stones were laid on the south-eastern approach to Blackfriars Bridge at about the same time in 1869, and after the York had been taken up and removed for some time the Victoria stone was apparently as good as when laid.

Regarding the strength of this stone, a recent issue of the *Universal Provider* states:—"The remarkable strength of this stone may be judged from the latest test, which proved that a crushing weight of 8321 lb. to one cubic inch was necessary before the Victoria stone succumbed to pressure, and the tensile strain is stated to have reached 1310 lb. per square inch. It is said to absolutely defy climatic changes, the atmosphere having no deteriorating effect upon it." The ordinary sizes in which these paving flags are manufactured for paving are as follows:—

2in. thick.	3in. thick.
3ft. 0in. × 2ft. 0in.	2ft. 0in. × 2ft. 0in.
2ft. 6in. × 2ft. 0in.	2ft. 6in. × 2ft. 0in.
2ft. 0in. × 2ft. 0in.	3ft. 0in. × 2ft. 0in.
1ft. 6in. × 2ft. 0in.	3ft. 6in. × 2ft. 0in.
	4ft. 0in. × 2ft. 0in.

Imperial Stone is another excellent paving material, and has many advantages over the natural stone. Its appearance is equal to that of selected natural stone, and it is easily laid, being beautifully squared in working, or it can be cut to any desired shape if necessary. This stone is manufactured from finely broken granite and Portland cement, in the proportion of 3 of the former to 1 of cement. These materials are thoroughly incorporated, which is effected by machinery, and a small quantity of water is added whilst the machinery continues in operation. Upon this being well mixed, the material is placed into proper moulds. These moulds and their contents are afterwards placed on a machine giving a rapid jolting motion, called the "trembler," which produces a uniform slab of non-porous and dense stone. Upon the slabs being so treated, they are allowed to remain in their moulds for two days, after which they are exposed to the air for about a week. This being done, they are indurated by being immersed for a further week or more in a solution made from flint and caustic soda. On the completion of the slabs, they are exposed to the air for some months to season before use.

Adamant Paving Stone, which has been much employed as a paving material for footpaths, is an artificial stone composed of finely-crushed Aberdeen granite and Portland cement. No chemical process is adopted in the manufacture of this stone, but each slab whilst in the mould is subjected to hydraulic pressure, which causes the block to become exceedingly dense and free from air cavities and moisture. The blocks being thus freed from moisture, soon become dry and ready for use. The ordinary sizes in which this flagging is manufactured are:—3ft. by 2ft., 2ft. 6in. by 2ft., 2ft. by 2ft., each 2in. thick.

Hard York Patent Stone.—This stone is better known under its registered name, "Non-slip" stone, designating that it possesses the special feature of being non-slippery. This stone as a paving material has been extensively used in London and elsewhere with much success. Hard York stone is used in its manufacture, and as in the case of the Adamant stone, the Non-slip slabs are subjected to great hydraulic pressure, which solidifies the material and frees it from moisture.

Elliott's Patent Stone, manufactured by the South of England Stone Company, has been used for paving of footpaths in the district of its manufacture. The materials employed in the manufacture of this stone are Kent ragstone, dust, Portland spalls, rock siftings, Portland cement, &c. The surfaces of the slabs are made with much care, although no pressure is employed, the process of finishing gives an unusually smooth and clean appearance to the slab; the material is, further, tinted to the colour of natural stone, yellow ochre and red oxide being used for that purpose. These slabs when laid have a very pleasing appearance, and are as smooth as rubbed Portland, and yet non-slippery. The wearing qualities appear to be good—some that the author laid five years ago are giving much satisfaction. These paving slabs are well matured before being sent from the works, a large quantity being always kept in stock for the purpose of seasoning. The sizes stocked are 24in., 27in., and 30in. wide; 2in. and 2½in. thick; and almost every length from 12in. to 36in. Approximate weight, 100ft. super 2in. = 1 ton; 80ft. super 2½in. = 1 ton.

Patent Granite Plumbic Paving is a paving material consisting of granite siftings and Portland cement, into which is introduced a quantity of small particles of lead. The manufacturers of this paving material—which can be either laid in slabs or *in situ*—claim that the introduction of lead "has the effect of rendering the surface always non-slippery." The following particulars are given by Messrs. W. Garstin and Sons, the patentees, as to the wear-resisting properties of lead as compared with stone:—"On a first consideration it may appear that lead, being of a softer nature than either granite or concrete, will not hold its own so far as wear is concerned against these materials. Further investigation will, however, show that this is not the case, for lead will not disintegrate and "carry," while both concrete and granite, with constant traffic, do to a certain extent wear, and the disintegrated particles are carried away. Therefore, instead of the lead wearing lower than the harder material, it has the reverse effect; each piece being in a cavity, cannot get away, therefore the elevated portions are beaten down by the traffic and extended over the wearing surface. Thus the introduction of lead into concrete, granite, or other materials improves rather than deteriorates its quality and durability, while it increases its value as a safe foothold."

CHAPTER XXII.

CONCRETE LAID IN SITU.

THIS description of footpath has become a favourite amongst many Authorities, especially in districts where the carriage on other paving materials adds considerably to their cost. Concrete laid *in situ* has in many cases been a failure, owing to its becoming much cracked, and in some instances raised off its foundation, from the effects of changes of temperature. This failure is the result of the concrete being laid over too large an area without allowing for expansion. This work is now usually executed in sections of 3ft. wide at a time, so laid that the paving should be divided up into separate slabs or squares. Pavements of this description, properly laid with suitable materials, have a particularly good appearance; they wear well, and are usually inexpensive. The chief drawback to their employment is that the pavement cannot be readily removed for the purpose of repairing or laying in pipes or cables.

The method of laying this description of pavement is as follows:—The ground having been excavated to the depth required, a foundation of 3in. of hard core should be made, consisting of broken bricks or stone of about 2in. gauge; over this a thin coating of fine siftings or cinders should be spread to fill up the interstices. This foundation should then be consolidated by rolling, on the completion of which a thin coating of clean sand to fill up any holes that may remain should be rolled in. In the case of good solid ground the thickness of the foundation may be reduced. Upon this foundation proper screeds should be placed, 2ft. 6in. to 3ft. apart, so as to divide the length up into equal widths. These screeds are strips of wood about 1in. in thickness, in depth equal to the finished thickness of the path; the wood must be well seasoned, and all sides planed smooth. Cross pieces should also be made to split up each section into the required number of divisions. All cross divisions should be made to break joint, and the lines should be carefully set out before the pavement is commenced. The pavement should be laid in two separate layers of concrete, and each sub-division laid and completed separately and allowed to set before the adjoining ones are commenced. The lower layer should consist of a thickness of concrete made in the proportion of 1 part of Portland cement, 2 parts of clean granite or other sharp stone broken to a $\frac{3}{4}$in. gauge, 1 part of clean, fine siftings, and one part of clean, sharp sand, which, after being thoroughly mixed when dry, and afterwards with water, should be rammed into the spaces between the screeds, leaving 1in. space for topping. Before this layer

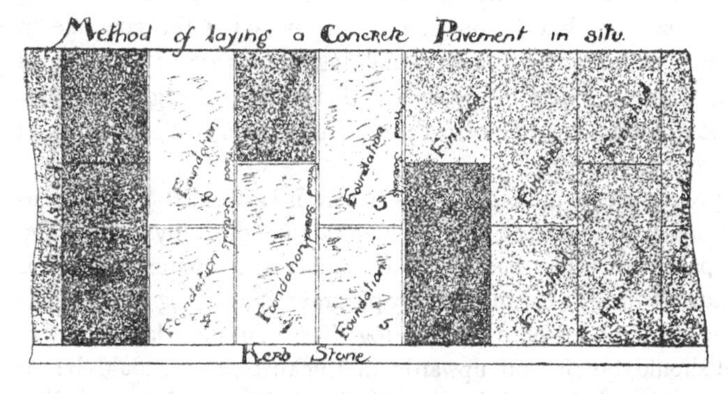

Fig. 45.

becomes set the topping or finishing coat must be laid, consisting of concrete made in the proportions of 1 part of the best finely ground Portland cement to 2 parts of fine well washed granite siftings of from $\frac{1}{4}$in. to $\frac{1}{16}$in. gauge. This finishing coat should be carefully rammed, and afterwards trowelled to a smooth surface, perfectly level with the edge of the screed on each side.

Upon these sub-divisions becoming set, the wooden cross piece forming the sub-division may be removed and a strip of brown paper laid against the section of the finished portion for the next division to abut against. In their turn the screeds dividing up the length of the pavement may be removed and strips of brown paper introduced for the next portion when laid to abut against. All wood is by this means removed from the pavement as the work proceeds, which is most important, as wood, when left in the pavement—as is sometimes the case—rots and disfigures the path before it is worn out. On completion of each division of pavement the surface should be sanded and kept slightly damp, so that the sun should not dry the concrete too quickly. (Fig. 45.)

Mr. Q. A. Gillmore, in his work on "Roads, Streets, and Pavements," gives the following carefully detailed particulars of the manner adopted in America for the construction of footways of concrete laid *in situ* :—

"Concrete footpaths should be laid upon a form of well-compacted sand or fine gravel, or a mixture of sand, gravel, and loam. The natural soil, if sufficiently porous to provide thorough subdrainage, will answer.

"It is not usual to attempt to guard entirely against the lifting effects of frost, but to provide for it by laying the concrete in squares or rectangles, each containing from 12 to 16 superficial feet, which will yield to upheaval individually like flagging stones, without breaking, and without producing extensive disturbance in the general surface.

"When a case arises, however, where it is deemed necessary to prevent any movement whatever, it can be done by underlaying the pavement with a bed of broken stone or a mixture of broken stone and gravel, or with ordinary pit gravel containing just enough of detritus and loam to bind it together. In high latitudes this bed should be 1ft. and upwards in thickness, and should be so thoroughly sub-drained that it will always be free from standing water. It is formed in the usual manner of making broken stone or gravel roads already described, and finished off on top with a layer of sand or fine gravel, about 1in. in depth, for the concrete to rest upon.

"The concrete should not be less than 3½in. and need rarely exceed 4in. to 4½in. in thickness. The upper surface to the depth of ½in. should be composed of hydraulic cement and sand only. Portlant cement is best for this top layer. For the rest any natural American cement of standard quality will answer. The following proportions are recommended for this bottom layer :—

Rosendale or other American cement	1 measure
Clean, sharp sand...	2½ "
Stone and gravel	5 "

"It is mixed from time to time as required for use, and is compacted with an iron-shod rammer in a single layer to a thickness less by ½in. than that of the required pavement. As soon as this is done, and before the cement has had time to set, the surface is roughened by scratching, and the top layer, composed of 1 volume of Portland cement and 2 to 2½ volumes of clean, fine sand, is spread over it to a uniform thickness of about 1½in. and then compacted by rather light blows with an iron-shod rammer. By this means its thickness is diminished to ½in. It is then smoothed off and polished with a mason's trowel, and covered up with hay, grass, or other suitable material to protect it from the rays of the sun and prevent it drying too rapidly.

"It should be kept damp and thus protected for at least ten days, and longer if circumstances will permit; and even after it is open for traffic a layer of damp sand should be kept upon it for two or three weeks, to prevent wear while tender. At the end of one month from the date of laying, the Portland cement mixture forming the top surface will have attained nearly one-half its ultimate strength and hardness, and may then be subjected to use by foot passengers without injury. The rammers for compacting the concrete should weigh from 12 lb. to 20 lb., those used on the surface layer from 10 lb. to 12 lb. They are made by attaching rectangular blocks of hard wood shod with iron to wood handles about 3ft. long, and are plied in an upright position. Certain precautions are necessary in mixing and ramming the materials in order to secure the best results. Especial care should be taken to avoid the use of too much water in the manipulation. The mass of concrete, when ready for use, should appear quite incoherent and not wet and plastic, containing water, however, in such quantities that a thorough ramming with repeated, though not hard, blows will produce a thin film of moisture upon the surface under the rammer, without causing in the mass a gelatinous or quicksand moton."

CHAPTER XXIII.

ARTIFICIAL STONE PLANTS.

Artificial Stone Plants.—Town Councils sometimes manufacture their own artificial flagstones, utilising the waste materials at their quarries or the slag from their dust destructor for this purpose. In order that this work may be satisfactorily and economically carried out, some description of artificial stone plant is necessary. By the use of hydraulic artificial stone presses, much waste material can be turned to account by being converted into good paving slabs. Paving slabs, even when made of the best granite chippings and cement, are only suitable for light or medium traffic, unless prepared by some patent process or compressed by hydraulic machinery. The hydraulic artificial stone plants manufactured by Messrs. Fielding and Platt, Limited, with their recent improvements (Fig. 46), are amongst the best that have come before the author's observation. This firm of engineers claim to be the oldest makers of special hydraulic presses for the production of concrete flags, blocks, &c., the first presses ever put down in this country being constructed by them about ten years ago at Shep, Westmoreland, at Aberdeen, and at Greenwich respectively. With a Fielding and Platt press thirty slabs can be thoroughly compressed in an hour, with the employment of one man, who is able to do the whole of the manual labour connected with the working of the press, together with that of discharging and filling the moulds. It is necessary, of course, to employ the usual men in mixing the concrete, and in carrying away and stacking the slabs on completion.

A greater number of slabs an hour could be produced by this plant, but as the quality of a slab is improved by being subjected to a prolonged rather than a minimised duration of pressure, it is better, in order to ensure durability, to adopt the slower and surer process. The slabs, after being pressed, are automatically transferred from the press to the discharge table by means of a pneumatic lift. This is an ingenious feature, as it avoids the necessity of handling, and the risk of damaging the newly-moulded concrete.

Mr. A. H. Mountain, Surveyor to the Withington Urban District Council, where one of Messrs. Fielding and Platt's im-

Fig. 46.

proved plants has recently been laid down, speaks very highly of it, and says he is able to turn out paving slabs at a cost of under 2s. per superficial yard.

Berry's patent artificial flag machinery is manufactured by the well-known firm of Messrs. Henry Berry and Co., Limited, Leeds. This firm of hydraulic engineers has given a great deal of attention to the production of a hydraulic press, consisting of the very best and improved machinery, especially designed to produce high-class artificial paving flags with the greatest possible economy. In this machine a heavy revolving table, made in the form of a cross, is a great feature in its favour. Two rams are provided in connection with this table, one for the purpose of pressing the flags and the other for removing the flags from the moulds. Fig. 47 shows this table and the fixed bed upon which it revolves, the latter being planed to a smooth and perfectly level surface. Each

Fig. 47.

arm of the revolving table is provided with a suitable opening for the reception of a mould, the latter being bolted to this table as shown. A closely-fitting removable plate is placed in each mould, upon which the concrete is spread. Upon the mould being filled with concrete, a quarter-turn is given to the revolving table, which carries the charged mould to the position next to the press. The concrete is here levelled by a second man or boy in attendance, and a second quarter-turn carries the filled and levelled mould to the third position under the powerful ram. This ram is then set in motion, and, to commence with, applies a pressure of 500 lb.

per square inch, which pressure is increased by intensifying to about 3500 lb. Upon this pressure being removed another quarter-turn is given to the table, which carries the flag to the fourth position under the second or smaller ram, which forces the flag and loose plate out of the mould, through the opening in the table on to a trolley underneath, and in like manner the work is repeated. All four operations are in progress at the same time, four men or boys being employed at the machine, each taking his particular part in the work. By this means there is absolutely no time wasted, and the manufacturers claim that high-class flags can be made at the rate of one a minute. This machine is adapted for making flags of any sizes from 2ft. by 2ft. to 2ft. 6in. to 3ft., and a comparatively small space is required for the erection of the plant. Messrs. Berry and Co., beside their patent machine already described, are manufacturers of the ordinary single and double mould flag presses; however, there can be no doubt that this new arrangement is far in advance of the previous types.

Messrs. C. and A. Musker, Limited, Liverpool, manufacture very excellent concrete flag-pressing machinery, which has the highest testimony of leading engineers, who have adopted their installations for making concrete paving slabs from cement and destructor clinker, or crusher rock. This firm of engineers makes three types of press plants, of which the "C" type press is one of the most powerful and strongly constructed machines invented for the purpose intended. The machine has been designed on the most practical lines, to produce with as little loss of time as possible a pressure of 200 tons on each flag. The amount of power required for producing this great pressure is 2 horse-power. The flags made by this machine, under a pressure of 200 tons, are of the highest quality, and as they can be rapidly made, their cost of manufacture is consequently low.

CHAPTER XXIV.

ASPHALT.

ASPHALT, besides being used for carriage-ways, is largely employed as a paving material for footways; it makes an excellent footpath, being durable, non-slippery, expeditiously laid, pleasant to walk upon, and is not glaring to the eyes. Asphalt makes a comparatively inexpensive footpath; a path laid about 1in. thick of good asphalt will wear from ten to twenty years, according to traffic, and when taken up for renewal the old material may be melted up again and relaid. Asphalt should be laid on a perfectly solid foundation, all soft materials being removed and good stone or gravel substituted in its place, thoroughly consolidated by rolling. The surface of the foundation should be graded to a depth below the kerb equal to the thickness intended for the asphalt, or if a layer of concrete is to be laid to carry the asphalt—which is recommended in all cases where there is much traffic—an extra depth must be taken into account.

The concrete used in the foundation should be 4in. thick and consist of 1 of Portland cement to 5 of clean sharp stone, with only sufficient fine stuff mixed to fill up the interstices. This, being properly mixed, should be laid in sections of 3ft. in width across the path. The joints should be made in the same manner as that for concrete paving laid *in situ*. No asphalt should be laid on the concrete until the latter is perfectly dry. Foundations of concrete are liable to expand and slightly crack the asphalt. This is the case more with compressed asphalt than with mastic, owing to a lack of elasticity in the former; but these slight cracks almost immediately close up again in busy streets.

In compressed asphalt the heated powder—consisting of the natural rock ground very fine—is brought to the site of the work in iron covered vans, and is then laid and raked over the surface of the foundation already prepared, and rammed with hot pelons to the thickness required. It is then ironed to a smooth surface by means of a hot ironing tool, and whilst the material is still warm, is rolled with a heavy hand roller. The finished thickness of this material is about 1in., and sometimes 1½in.

Mastic Asphalt is more commonly used upon footways than compressed asphalt. This raw rock should be finely ground, well impregnated, and of a regular brown colour. Great care must be taken that no inferior materials are substituted in the place of bituminous rock, such as pitch, or a manufactured imitation by the use of chalk, &c. The natural rock is broken up into small pieces, and melted in proportion of 15 of mastic to 1 of bitumen, in hermetically closed boilers. When the mastic is properly melted a quantity of clean fine sand or fine gravel is added, equal in proportion to two-thirds of the foregoing mixture. The mixture of mastic bitumen and sand being well heated together is run out in sections on to the footpath of about 8ft. wide, and evenly spread with a hand float, and finished by rubbing fine sand into the surface. Mastic asphalt is sent to different parts of the country, prepared at the works in cakes, which need only to be melted down in ordinary cauldrons and laid in the manner described. A small additional quantity of bitumen is added in this case to make up for loss by evaporation. Mastic asphalt liquefies at a temperature of about 300 deg. Fah.

Asphaltic Limestone Concrete.—This paving material for footpaths is manufactured by the Asphaltic Limestone Concrete Company, Limited, and is composed of Derbyshire limestone cemented together with a bituminous mixture. Pavements laid with this material are satisfactory, and have a clean appearance, as it wears white, and yet possesses the advantages of an asphalt in being slightly elastic, non-absorptive, non-slippery, easily laid, and inexpensive.

CHAPTER XXV.

TAR CONCRETE AND OTHER PAVEMENTS.

TAR concrete pavements are very extensively used, and when well made and laid, make one of the cheapest and pleasantest foot-paving materials. The material can be prepared ready for use by any Corporation's workmen, using the chippings from the stone-breaker, sifted to the sizes required for the fine and coarse material. The stone chippings should be of about ¼in. gauge for the topping material and about ¾in. gauge for bottoming. These sizes should be prepared separately, but in the same manner. The stone, which must be free from dirt, should be placed over a coke fire, on a large iron plate, of about 20 superficial feet, and well heated. The cementing material, consisting of tar, pitch, and creosote oil, is heated in cauldrons, and both the stone and tar mixture are mixed together whilst in a heated state. Each hundred-weight of stone requires about 1 gallon of tar, 2½ lb. of pitch and ¾ pint of creosote oil.

Gas tar alone may be used, and gives very satisfactory results when properly boiled; but if not sufficiently boiled, the pavement will become soft and sticky during hot weather. The water, as it rises to the top of the boiling tar, should be removed by means of a flat tin shovel.

The best materials for tarred concrete are Kentish ragstone, Derbyshire limestones, and other porous stones. An excellent material is supplied ready for use, or laid if required, by Messrs. Bensteads and Messrs. Constable, from Kentish ragstone, out of which a good path can be made at from 2s. to 2s. 6d. per yard super, within reasonable distance from Maidstone. Granites and other similar stones make a very serviceable paving material, but do not roll so close as the softer stones. The material should be prepared ready for use in the winter months, and laid in the spring when the weather is fine and warm. Any work done during wet weather will be a failure. When hard stone is used, the materials are more solidly compressed when laid in hot weather; but in this case a thick coating of sand must be immediately rolled over the finished surface to keep off the rays of the sun for a week or two. The foundation should consist of 3in. of coarse tarred stone properly consolidated by rolling and 1in. of the finer material. The "topping" coat should be spread on and rolled with a light roller weighing about 3 cwt., and, whilst rolling, the path should be gradually added to till about ¼in. above the kerb, and with the slope required. Some crushed marble or spar may be sprinkled over this surface and lightly rolled in, which gives the path a nice appearance when finished. Upon this being done, and the surface better able to stand the men's feet, a heavier roller of about half a ton in weight should be applied, and the rolling continued until the path is quite consolidated.

A greater area of path than can easily be completed in one day's work should not be commenced at a time, as each portion laid needs to be completely consolidated by rolling on the same day that the material is laid. During the whole of the process of consolidating the face of the roller must be kept damp, which may be effected by occasionally applying a wet cloth. Care must be taken that no water falls on to the paving material whilst damping the roller.

The surface of tar concrete paths should be brushed over with hot well-boiled tar once every two or three years after laying, to preserve the path, and a coating of fine sand or calcined shells spread over the coating of tar, to be trod in by pedestrians. The life of tar concrete periodically "dressed" is very great. One of the principal promenades at Margate was laid with this material seventeen years ago, and appears to be as good now as the day when opened to the public.

Tar Pavements.—The pavement already described as "tar concrete" is usually known as tar pavement, but the author, having had much experience of a surface made of a tar preparation only, has departed from the usual terms. Tar pavements are the cheapest foot pavements that can be made for situations where the traffic is light, and have been very extensively used at Margate, and more recently at Penzance. These paths consist of coating the surface of an evenly-graded and dry path made of fine gravel and clay, or, better still, of road scrapings, with a thick hot tar preparation, consisting of tar boiled in a cauldron for about three hours, and with the addition just previous to use of a small proportion of coal tar pitch, or pure bitumen. This mixture is spread over the surface with tar brushes, and whilst hot is well sanded with sand or fine shell and rolled with a light roller. The path requires to be twice coated in this manner in the first three months, and one ordinary tar dressing every twelve or eighteen months afterwards. The cost of this work at Margate averaged 1d. per superficial yard per annum, and many much-used promenades are prepared in this way. At Penzance the first cost of this work (two coatings) for an area of 20,000 yards superficial was 4d. per yard, including forming original surface of gravelled pavement, and the annual cost is 1¼d. per yard superficial.

Brick Pavements.—Bricks of buff, red, and blue colour are largely used for footways. The best buff-colour bricks, diapered to a suitable pattern for foothold—Fig. 48—make an excellent footway pavement equally on steep and flat gradients; they also have a very pleasing appearance when carefully laid. Red and blue-coloured bricks are also much used in some localities, and make a sound pavement. The brick are laid flat. A foundation of concrete is usually laid for superior pavements, with a layer of dry sand spread and rolled over its surface, on to which the bricks are bedded and levelled. In the place of concrete a bed is often prepared composed of macadam stone or clinker ashes, well rolled, with a layer of sand over the surface, as a bed for the bricks to be laid upon. The bricks should be laid in parallel rows across the path, at right angles with the kerb. They should be set in cement mortar, and the joints well filled and finished as close as possible. Bricks for footpaths are generally 8½in. by 4in. by 2¼in. in size, and should be well burned, tough, and vitrified, free from flaws or cracks, and of regular quality, colour, and shape. Ordinary red bricks are not satisfactory for footpaths, as they are soft and wear unevenly, much more so than is the case with vitrified bricks. The life of a properly-constructed brick foot pavement is long. The author has seen some blue bricks that have been in constant use for foot traffic in the Midlands for nearly twenty years, with scarcely the appearance of having had any wear at all. In some favoured situations these bricks should be equal to fifty years. Blue Staffordshire and Buckley bricks, although both very durable, do not make a path of good appearance, and are not recommended for good class streets. Buff bricks, such as those manufactured by Messrs. Candy and Co.,

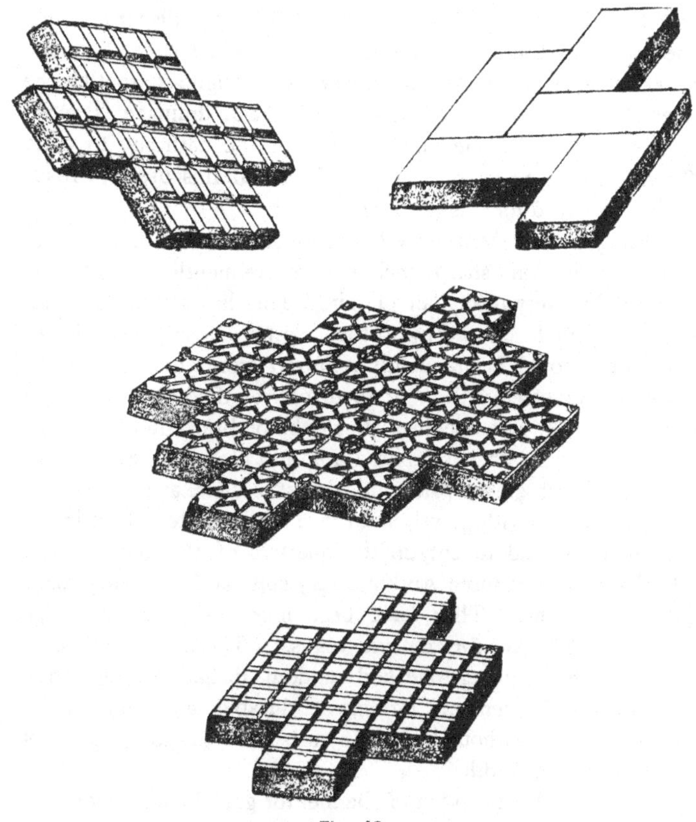

Fig. 48.

and Messrs. Hexter, Humpherson and Co., make both a durable and pleasant path, quite suitable for good residential streets.

Gravel is much used for footpaths for country roads, parks, and gardens. Gravel paths, unless well constructed and drained, are a source of constant expense and trouble, but when these essentials are efficiently attended to gravel forms a good and inexpensive pathway. Gravel paths require a foundation prepared with a layer of well-rolled macadam stone, over which should be placed 3in. or 4in. of gravel well rolled in layers while the material is damp. The traffic should be kept off until the work is dry. If the material bed is found to be sufficiently firm, the gravel may be placed on it direct and the foundation of stone dispensed with. The bed should always be formed with a sufficient camber, and the gravel spread on in equal thickness throughout, so that the water should run off into the channels. Sub drainage is sometimes necessary at one or each side of the pathway, to carry off water from springs, &c. Any of the sub-drains used for roads can be employed for this purpose.

Nothing is more destructive to gravel paths than surface water scours; paths on sharp inclines are frequently considerably damaged by every heavy fall of rain. This is difficult to avoid, but can be much reduced by a proper system of surface drainage. The water should be frequently removed from the channels by openings connected with a drain to convey the water away to a suitable outlet. The openings should be of brick similar to road gullies, provided with catch-pits, traps, and gratings. Small traps and gratings for gravel paths are almost useless, as they so soon become silted up with gravel, and cause the water to collect in the side channels, and to cut up the quarters of the gravel paths. Side channels of random work are very suitable for country roads and gravel paths. They have been much used at Tunbridge Wells and Margate for outside roads. There they are constructed with Kent ragstone laid random, dished out in section and grouted in cement, forming a durable and inexpensive channel with or without kerb. It costs from 9d. to 1s. per foot run, according to width.

Another good description of channel for gravel walks is concrete laid *in situ.* Tar concrete has also been used with success.

CHAPTER XXVI.

STREET RAILWAY TRACKS.

TRACKS are usually constructed of rails supported on chairs or sleepers, or in many cases laid directly upon the concrete foundation. The permanent ways are paved with granite or other stone setts, and not infrequently wood paving is introduced. The form of rail that appears most effectually to meet the requirements of a railway track, and also to interfere as little as possible with

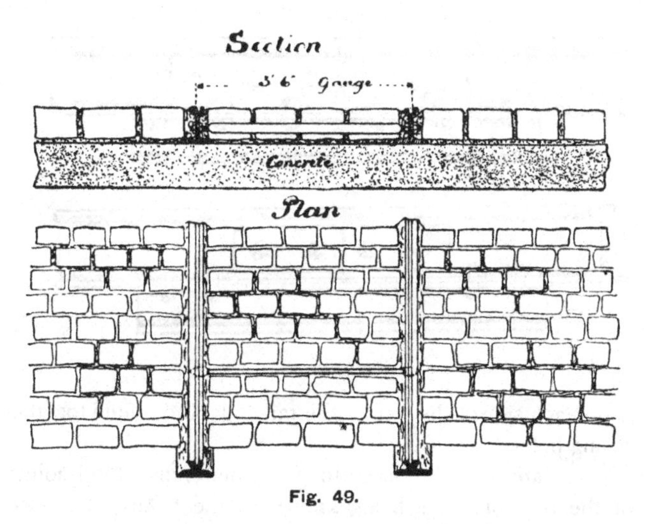

Fig. 49.

the ordinary street traffic, is the "girder rail." This rail is manufactured in the form of a web girder provided with a lower flange to be fastened down to 8in. by 4in. or 9in. by $4\frac{1}{2}$in. sleepers, or other means provided, whilst the head of the rail takes the form of a channel or groove for the wheel flange to travel in. This form of rail enables the paving to be kept flush with the rail, which is not the case with the use of the **T** rails.

Fig. 49 shows the system of construction usually adopted when sleepers are not used, the rails being bolted together by means of iron rods carried across the track. It is an advantage to fill in solid the spaces at the sides of the web of the rail with fine concrete or other packing, which will form a good face for the paving to butt against, and an additional support to the rail itself. It is of great importance that the joint between the rails should be as substantial as possible, failing which the rail will become "hog-backed," and a springing action will occur at the rail joints.

Fish-plates employed for jointing rails consists of iron plates

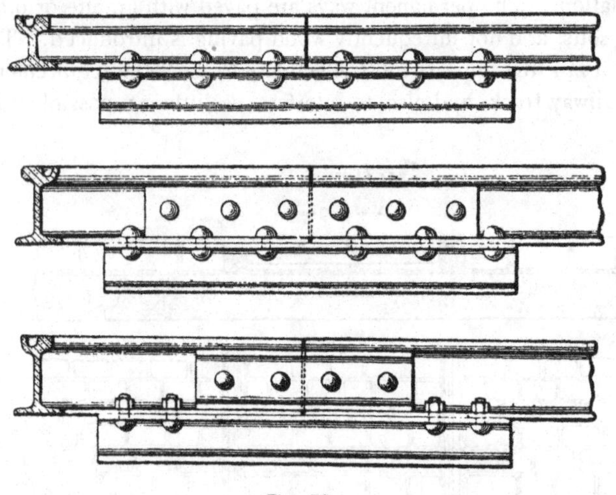

Fig. 50.

placed on each side of the web of the rail ends, and bolted together with strong bolts.

Sole plates are frequently used to strengthen rails at all joints, one of the best of these being known as the "Anchor" plate, patented by Messrs. Cooper and Howard-Smith. These sole-plates are embedded in concrete, and the rails are securely bolted down to them, forming a rigid and solid attachment, so desirable at the joints of all tram lines. The following illustrations show the patent "Anchor" plate fixed to rails, with and without the use of fish-plates—Fig. 50. The rails are secured to the sleepers by dog spikes and bolts, and on curves these should be used at frequent intervals. Small inspection boxes are sometimes provided at the ends of the rails, to facilitate access to the jointing bolts, so that these can be tightened up without removing the pavement.

Fig. 51 shows the form of rails used at places where stationary and movable points are required for passing places, &c.

Points and crossings made of Hadfield's patent manganese steel, or of best cast steel with manganese steel tongues, are to be recommended for durability.

The suitable paving of tracks is an important matter. Tracks should be laid with paved surfaces of the most durable nature

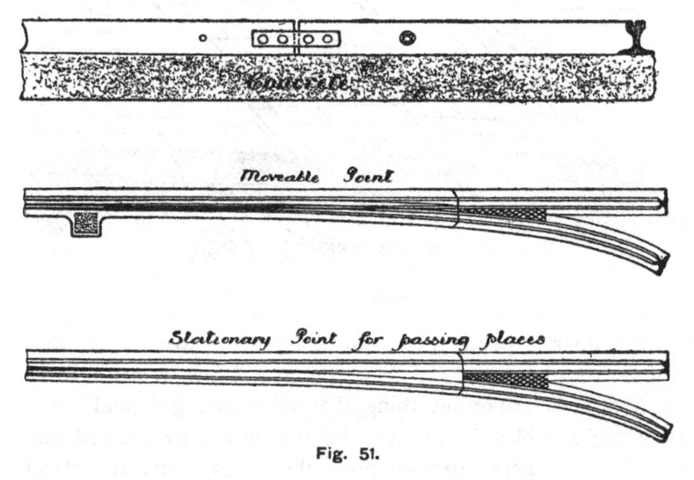

Fig. 51.

Granite sett paving is most commonly used for this purpose, and were it not for the disadvantage that vehicles passing over this paving create much clatter, it would be the best practice to pave all permanent ways with granite ; but as there is this objectionable drawback attached to granite setts, it is in many places advisable to employ hard wood or other suitable material in this work.

The question of expansion and contraction of the hard wood paving is one of the matters to guard against. Where the whole width of the road is not paved with wood, the outer edge, against which the macadam is to be rolled, can be successfully protected by a row of 4in. granite.

The Bingham patent paving should be found to answer well for the paving of trackways; so far as it has been tried, the method has given a great deal of satisfaction.

Another material, which will probably be much used in the future in tramcar tracks, is the sanitary block pavement, Fig. 52, which beside being a durable and non-slippery pavement for streets, can be adopted in tramways with the special advantage

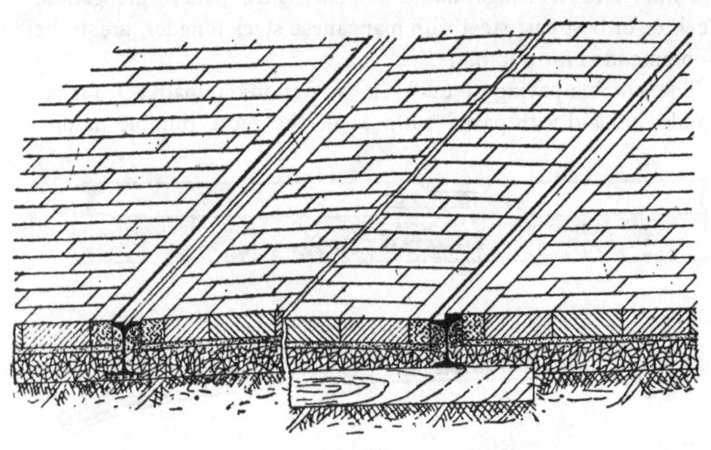

Fig. 52.

that it neither expands nor contracts under extremes of temperature.

It is a most important thing that all water, gas, and other service pipes, cables, &c.. should be thoroughly overhauled and renewed where necessary, wherever they pass under the street track, before the line is laid, or a lot of additional trouble and expense may follow in taking up the paving to rectify defects.

CHAPTER XXVII.

SCAVENGING OF STREETS.

ALL roads should be maintained in the highest state of cleanliness. The most successful roadmakers are most anxious to preserve their work in a complete state of cleanliness, so that good road-making and cleanliness should go hand in hand. Nothing stamps the character of a town and the class of its inhabitants more than the cleanly state of the roads and paths, without which there is the appearance of neglect on all sides, which is most noticeable to visitors accustomed to well-cared for streets. Cleanly roads and courts throughout a town play a very important part as regards the general healthiness and sanitary condition of the district. The cause of diarrhœa and some other diseases are known to be frequently due to germs or micro-organisms existing in the soil, which, when saturated with filth and organic matter grow in profusion.

Professor Tyndall has thrown much light on the injurious nature of street dust. He showed that the dusty street air is literally alive with the germs of the bacteria of putrefaction, whilst the air on mountain peaks in the Alps is quite free of such germs.

Besides these very undesirable conditions, there is the great inconvenience that pedestrians and the traffic generally suffer during wet weather, when efficient cleansing is neglected. Not only would this inconvenience be much diminished at the time of rainfall by the regular removal of the detritus and droppings, but also the period required for drying the surface would be much reduced. It thus also follows that injury is done to the road itself by remaining for a disproportionate period in a wet and wasting condition.

General Sir John F. Burgoyne's reference to cleansing, in his "Remarks on the Maintenance of Macadamised Roads," is still valuable as a guide in this matter. He says :—" With regard to the cleaning, it will be done chiefly (it may be hoped *entirely*) with the broom; it will perhaps be difficult to be executed amidst all the traffic, but not so much so as would appear, judging from the present state of the roads, because it is contemplated that at no time will there be more than a very slight quantity of dirt to remove, and that chiefly of the matter dropped upon the surface. It is impossible exactly to foresee how often each part will require to be swept over. . . . . It must be the endeavour to keep *perfectly clean* and free at all times from mud or dust whatever parts may be submitted to the operation, so that every wheel may roll over the hard, compact surface of the stone alone. . . . These streets were constantly kept perfectly clean, hard, and even, and the material was of a tough, good quality, the actual wear of the surface would be extremely small—less, no doubt, than *one inch* in the year, even in those most frequented; the wheels, in fact, would be running over a smooth pavement made of small materials."

Mr. H. Percy Boulnois, M.I.C.E., who has, for many years, been a weighty authority on cleanliness of towns, in his useful book, "The Municipal and Sanitary Engineers' Hand-book," says:—"In most towns it is necessary to cleanse its principal streets at least once a day, and this appears to be the practice of nearly all the ninety towns I have referred to; only seven of them, however, appear to have this operation repeated more frequently; in several towns the horse droppings, &c., are removed at once, under what is called the *orderly* system, and this is especially necessary in streets that are paved with such materials as wood paving, asphalt, or granite setts. The suburban streets of a town need only be cleansed once or twice a week, except in special cases of extremes of mud or snow. . . . ."

Mr. D. K. Clark, C.E., in Part II. of the "Construction of Roads and Streets," gives his valuable experience on this subject as follows:—"By far the greatest proportion of the detritus of macadamised roads consists of the worn material of the road; and the important principle was early revealed by experience, 'that the oftener that streets are cleansed the less is the mud which is created and removed, whilst the attendant expenses are by no means increased, and the roads are kept in a better state of preservation.' This principle is, besides, logically deducible rom the fact that the worn particles, if left on the surface, act as a grinding powder under the wheels and the horses' feet to reduce to similar powder the surface of the roads, and that the mud which is formed with the detritus, when rain falls upon and is mixed with it, operates like a sponge in retaining the moisture upon the surface. The upper crust, as well as the substratum, under these circumstances, become saturated with moisture, softened and 'rotten'—just as a gravel footpath, hard and solid in ordinary weather, becomes sodden and pulpy when it is lapped for a time by a covering of half-melted snow. The macadamised carriage-way, thus reduced, is exposed to rapid deterioration by the traffic, which increases in a highly accelerated ratio with the period during which the road is left uncleansed. The statistics of cleansing unanimously support these conclusions."

No further comment is necessary to show the great desirability of maintaining all streets in a state of thorough cleanliness, as being the surest means of conducing to the permanence of roadways, in securing health and comfort to the inhabitants, and the good appearance of the streets and town generally.

This condition of cleanliness should not be restricted to the front and most frequented streets, but should most decidedly be extended to all back roads, alleys, and courts. Such spaces existing in the back and poorer localities are, as a matter of fact —owing to the careless habits of this class of inhabitants, together with the generally narrow and confined areas which exclude the drying and purifying action of the sun and wind— the more needful of attention on sanitary grounds than many of the more favoured ones. This work in some towns receives a good deal of attention, but in others, where the Authorities are not so alert in sanitary matters, the duty is more or less neglected on the score of false economy.

It is an encouraging fact, when once systematic cleansing of back courts and lanes has been set on foot, how soon the inhabitants become educated up to a higher standard of cleanliness; so that the cost of such work, which at the commencement may have been rather heavy, in course of time becomes more than repaid indirectly by the improved habits of the people.

On the important question of scavenging of private courts and alleys, the author cannot quote any better authority than that of Mr. H. Percy Boulnois, C.E., already referred to in this chapter. He says:—"It is very questionable, however, whether the onus of cleansing private courts and alleys which are not repairable by the Urban Authority should be borne by them, although for the sake of the public health it is highly desirable that such work should be so undertaken. The great difficulty attached to this duty arises from the fact that these private courts and alleys are generally very badly paved—if paved at all—full of pits, where pools of stagnant mud and water collect; and even in the best cases the interstices between the pebbles, or other paving, are filled with filth, arising in great measure from the dirty habits of the people; and this filth it is found extremely difficult to dislodge. The remedy for this is to compel the owners of the abutting properties to have the courts and alleys properly paved with asphalt, or other equally impervious material, after which it would be easy for the Urban Authorities to cause them to be swept at least once a day, and flushed with water in the hot weather once a week; but in order to compel the owners to execute this very desirable work, it would be necessary to put the complicated machinery of Sec. 150 of the Public Health Act, 1875, or of the Private Street Works Act, 1892, in force, and the expense to the landlords would be in many cases very disproportionate to the value of their property. Out of the ninety towns to which reference has before been made, the Authorities of only nineteen of them cleanse the private courts and alleys in their jurisdiction, although for the sake of sanitation it is very desirable that such work should be so undertaken by them."

CHAPTER XXVIII.

CLEANSING.

THE methods employed for cleansing consists of hand labour either assisted by machinery or not.

The work is sometimes carried out from daybreak during the early hours in the morning, whilst in other cases the cleansing is done by night, and only the street orderly system for the removal of the horse droppings is continued during the day. In other cases the scavenging is done partly by night and partly by day. The method usually adopted is for the Authority to employ their own staff of workmen and team labour.

This is a system highly to be commended, and it would in most cases, upon full consideration, be found that the adoption of this course, instead of the plan of letting the work by contract, would be much to the advantage of the Authority.

For the removal of dust and mud from the roads surface sweeping and scraping are the usual processes employed. These processes must, to some extent, be carried out by hand; but the labour and cost is much reduced, and the work more speedily done, by the use of machine brooms and scrapers wherever possible. Bass brooms are generally used for sweeping, and are particularly suitable for macadamised roads, but for asphalt and wood paving rubber squeegees are much employed, and are very suitable in wet weather. With a bass broom a man can sweep an average of 120 yards lineal of roadway 25ft. wide in a day; the amount of work done, however, much depends on the condition of the road and the state of the weather. One man may be able to keep an area clean during dry weather which it would take two men to do during wet weather. It is best for the Authority to furnish all brooms necessary for sweeping, otherwise inferior work may be done by the use of unsuitable or badly worn articles. Brooms as soon as they become worn should be placed on one side as unsuitable for further roadwork, as their use is liable to injure the surface, and is not so well suited for expeditious work. Old broomheads, when not too badly worn, will realise a ready market amongst farmers, &c.

The author has invented a broom attachment clip road scraper. This attachment forms a very useful addition to the broom for cleansing purposes, as by its employment all substances can be readily removed from the road's surface, which, if accomplished

Fig. 53.

by the broom only, would mean much waste of time, additional wear to the broom, and the removal of quantities of small but serviceable material from the road.

The patent clip scraper is illustrated in use in Fig. 53. It consists of a steel blade rivetted to two steel straps, which pass round the head of the broom, and are firmly fixed in position by passing bolts through between the bass, connecting each strap thereby and tightening up by means of brass wing nuts.

The attachment is light; does not injure the broom in fixing, and is easily detached therefrom for reuse, upon the old broom-head becoming worn out.

A sweeping machine, with a set of revolving bass brushes, will sweep a track 6ft. wide and leave the dust or mud when finished at the side of the road ready to be gathered up and carted away. A sweeping machine drawn by a horse with one man in attendance will sweep an average of 120 yards lineal of muddy road 25ft. in width in one hour; or, in other words, a horse broom will sweep as much in one hour as one man will sweep in one day.

According to this experiment the cost of hand sweeping is about 3s. per 1000 yards super, as compared with 1s. (horse 6s. 6d., driver 3s. 6d., working 10 hours), the cost per 1000 yards super of horse broom sweeping, without taking into account the cost of brooms, &c., used for hand sweeping, or brooms and machine in the other case. The cost of a high-class machine sweeper is £33 10s.

Great improvements in combined sweeping and collecting machines have recently been effected, by means of which the mud or dust is automatically removed from the road whilst the machine is in motion. There is great economy in the employment of a successful machine of this character, as it dispenses with the usual and costly method of employing team and manual labour to collect the sweepings from the sides of the streets.

These new machines differ from the old types, in that the mud or dust cart used is detachable, and is not necessarily of limited size. A further feature in the machine, called the "Salus," is that the brushes are flexible, so that ordinary irregularities of the road are automatically compensated for by this. The "Salus" machine was invented only last year, and has since been adopted by the Municipal Authorities of Cologne and Elberfeld and elsewhere, and it is stated that one of these machines with two men and two horses took the place of twelve men, and taking into account the stoppages necessary for attaching and detaching the mud cart and other purposes, a track of two miles in length can be cleansed and collected per hour.

The desirability of sprinkling a dusty street with water before sweeping should recommend itself to all Authorities as effecting an improvement in the healthiness and comfort of such streets, whilst the necessity for the employment of a watering cart preceding a sweeping machine at such times as the mud is sticky (which without the aid of water would be ineffectually removed by the machine), is known to all who are experienced in this work. The inventor of this new sweeping machine, the "Salus," recognises these important desiderata, and provides an ingenious watering contrivance under the driver's seat, by which the road is sprinkled previous to its being swept, if found necessary.

A machine so fully equipped must needs be costly, but the manufacturers claim that the whole sum is recoverable in the course of one year by its economical working. A recent number of "The Street" states :—"Some interesting and minute tests have been made in Berlin with the 'Salus' sweeper by Mr. Th. Weyl, M.D., and lecturer in hygiene at the Charlottenburg High School. It appears from his report that without throwing up the least dust the machine cleared the road as well as if it had been done by hand; droppings, grit, stones of the size of a child's fist, nay, entire bricks, were dealt with indiscriminately by the broom. In his opinion, this apparatus meets every requirement of such a mechanical street sweeper. The only drawback appears to be its high initial cost of nearly £300, but this would indeed be a very poor objection to its introduction if the whole sum is recovered in the course of one year, as the manufacturers claim for their machine. We understand, moreover, that this price is subject to a material reduction as soon as a considerable demand is created, and this, it may be hoped, we shall not have to wait for too long. A German contemporary has calculated that in the case of an average scavenging district ordinarily served by two horse brooms and a night shift of twenty scavengers, the 'Salus' would effect an annual saving of £450 in wages, the staff (working at 3s. per head per day) being reduced by one half. Against this must be set the price of the machine, and the increased cost of maintenance, as compared with horse brooms. It is not likely, however, that the latter will swallow the balance of the saving, so that even in the first year after adoption a clear profit can be made, in addition to securing greatly improved sanitation."

Another ingenious machine for cleansing and automatically collecting the street sweepings or scrapings was invented by Mr. F. J. Scott a few months ago. This invention has for its object the provision of "an apparatus or machine which shall effect the cleansing of public streets and other places with considerably greater economy and expedition than has hitherto been possible with the means now employed." The mud or dust cart in this machine is detachable, and by a bucket or screw elevator the sweepings or scrapings are delivered into a wagon trailing after the cleansing machine.

The machine can either be fitted with scrapers or brooms, the former being constructed partly of india-rubber and partly of wood or metal, and, in two or more parts, suitably jointed together to enable the scraper to yield whilst passing over obstacles on the road.

Scraping.—Hand scrapers are useful substitutes for brooms for cleansing muddy roads. They consist of a steel scraper about

Fig. 53.

18in. wide and 6in. deep, with returned ends to retain the mud; this scraper is attached to a long handle rather thicker than that of a broom handle. Hand scraping machines are often used. These machines—Fig. 53—are made in widths of from 2ft. 8in. to 4ft., fitted with from eight to twelve shoes or scrapers; each shoe is 4in. broad, hinged to a horizontal bar, and pressed into position by a separate spring. The whole is mounted on two wrought iron wheels 2ft. in diameter. Most expeditious work can be performed by the employment of this machine. It takes but one man to work it and one to sweep the mud into heaps ready for its removal by a mud cart. The use of the hand-scraping machine affords one of the most economical methods of cleansing roads, besides being three or four times as quick, as compared with hand sweeping or scraping. The machines are light and inexpensive, their cost ranging from £3 to £6 each.

Horse Machine Scrapers.—In this form of scraper, as in the horse broom, the shoes are mounted on a frame in such a direction that the mud is delivered at the side of the machine when in motion. These machines are suitable for use when the mud is too thick and sticky for the sweeping machine. As in the hand scraping machine, all the shoes work independently of each other, and are fitted with springs, so that they adjust themselves to any uneven parts of the road without causing injury to its surface.

These machines are 6ft. in width, and cost from twelve to sixteen guineas each.

The question of the expeditious collection and removal of street sweepings is one of much importance. During wet weather this item is a large one, requiring additional horses and carts to those usually employed in dry weather. No sweepings should be allowed to remain after they are swept up at the sides of the road longer than an hour in wet weather, and less time in dry. Sufficient carts should be in attendance to pick up the detritus as soon as possible after it is deposited, and additional carts must be allowed for according to the distance that it has to be removed for disposal. Minor depôts in central positions of large towns are suitable for the purpose of temporarily accommodating the sweepings during the busiest times, to be removed at more convenient hours after the streets are cleansed. In this case it is often advisable to tip the mud from the mud cart into a large mud wagon or motor vehicle. For the purpose of tipping into the wagon a raised tipping platform would be necessary.

A lot of labour, room, and expense may be saved by this method, as liquid mud cannot be temporarily deposited in a yard without spreading over a large area. Streets paved with impervious materials make a considerable difference in the cost of team work for cleansing as compared with macadamised roads. This account will invariably be found to decrease in the ratio that the area of impervious roadways is increased.

Disposal of Sweepings.—It is often a difficult problem to decide how and where to dispose of the street sweepings. In smaller towns there is usually a ready market for the horse droppings for agricultural and garden purposes, but in large centres this is a more serious matter, and when combined with the street detritus, which generally amounts to thousands of loads per annum, necessitates some costly means of disposal.

In some seaside towns and others situate on tidal rivers it is often carried to sea and deposited in deep water; in other towns it is destroyed by means of the refuse destructor, which is by far the most sanitary method hitherto adopted; whilst occasionally, in some towns, advantage is taken of any inexpensive method, irrespective of sanitation or after results, such as the filling up of low-lying lands, old wells, or other disused shafts, which practice is most strongly to be condemned.

CHAPTER XXIX.

SNOW.

THE expeditious removal of snow from the streets is always a more or less troublesome business.

The removal of a big fall of snow entails many difficulties, and is an expensive undertaking.

In many towns the additional horses, carts, and labour that would be necessary for clearing the streets of a heavy fall of snow are engaged beforehand, so that they are at all times prepared in anticipation for the extra work; but in towns where this provision is not made, in the case of an unexpected fall of snow, a lot of time is sometimes lost in collecting the extra labour and horses necessary for its removal. When the fall occurs in the daytime arrangements can usually be made during the time it continues, as it is practically useless to remove the snow until the fall has ceased.

The removal of snow can be let by contract at a price per inch depth of snow fallen. This is the system adopted at Milan, where it has been found to answer exceedingly well. In this case the city is divided into small districts, and each district is allotted to a contractor. The average depth of snowfall in each district is determined from the depth of snow that has fallen on to several flat-capped stone posts, fixed in open spaces for this purpose. The contractors in this case are bound to find all labour, horses, and carts, whilst the necessary implements, such as shovels, scrapers, brooms, &c., are furnished by the city.

Snow Plough.—The snow plough is a very useful implement for the removal of the fresh fallen snow from the roadway to the roadsides. The snow plough should be used before the snow is consolidated by the traffic. Snow ploughs are generally made of wood, so designed that their weight can be added to if necessary by loading with stones, &c. It usually requires two or more rough-shod horses to draw an ordinary 6ft. plough. Care must be taken that the road surface is not ploughed up along with the snow by the use of unsuitable appliances.

Salt has been employed to assist in the removal of snow from the footpaths.

The following is from a report by the Superintendent of the Scavenging Department of Liverpool, referring to the use of salt :—"The only way to compass the removal of snow from the footwalks of the principal thoroughfares within a comparatively short time is by sprinkling them with salt, such as is commonly used for agricultural purposes. It is certain that, unaided by the salt, a sufficient number of men cannot be procured for the emergency of clearing snow from the footways of the most important thoroughfares. It has been stated by medical authorities that the application of salt to snow is detrimental to the health of people who have to walk through the slush produced by the mixture, and that the excessive cooling of the air surrounding the places where the application has been made is injurious to delicate persons. It, therefore, seems that the application of salt to snow should not be undertaken during the daytime, but should be commenced not before 11 p.m., nor continued after 6 a.m., and that only such an area of footwalks should be so treated on any one night as the available staff of men can clear by an early hour the following morning. To sweep snow from the footwalks whilst the fall of snow continues, and especially during business hours, appears to be wasteful and futile, and to apply salt during the same period may be held to be injurious to health. That the snow of an ordinary fall can be removed from the footwalks by an application of salt an hour or so before they are scraped is an ascertained fact, except at least where a moderately severe frost has preceded, accompanied, or followed the snowfall, or when the snow has drifted into extensive accumulations. Were it not for the danger to health by excessive cooling of the air, and for the expense attending the operation, all the impervious pavements could be cleared of snow (unless the fall was a heavy one) in a comparatively short time by a liberal application of salt and the employment of the horse-sweeping machines as soon as the snow has become sufficiently softened to admit of their use."

The onus of the removal of snow from the footpaths in many

towns falls on the occupiers of the premises abutting the street, which naturally relieves the Town Authorities of a lot of work, and much time can be saved thereby.

The disposal of snow after it has been removed from the street is often a question requiring much consideration. In seaside towns the snow can generally be tipped into the sea and disposed of in that manner; whilst in towns where there is a suitable river within easy distance, it could be carted to the river and there deposited. If neither of these courses is available, then advantage must be taken of any convenient open spaces, fields, or parks, where the snow can be conveniently heaped up with as little

Fig. 54.

damage as possible to any roads, paths, or grass. Both heat and steam have been tried for melting the snow as it leaves the cart, but neither process has been widely adopted, and, so far, successful and economical results are extremely doubtful; but in these days, wherein the refuse destructors are being brought so near perfection, a snow-melting addition may be looked forward to, and would prove a valuable help in this matter.

It is sometimes necessary to sand a street after a layer of snow has been removed, and occasionally during frosty weather. This work is more expeditiously performed by a cart fitted with a rotary sanding machine than by any other means. Fig. 54 illustrates a special cart manufactured by Messrs. Wm. Glover and Sons, Warwick, fitted with Willacy's rotary machine. The cart being made "hopper" shaped, it is quite self-emptying, requiring the attendance of the driver only. A width of 35ft. can be uniformly sanded by this machine.

CHAPTER XXX.

ROAD WATERING AND WASHING.

ROADS should be sprinkled with water as frequently as may be necessary for the purpose of laying the dust. Watering further supplies the need of moisture to the surface, so essential in dry weather to ensure the best wearing results of macadamised roads, and it also freshens and cools the air in the summertime. It is also desirable at times to wash paved streets with a stream of water, in which case it is advisable to remove the dust as much as possible from the surface previous to the application of any large quantity of water, or trouble may sooner or later ensue from the grit that would be carried into the sewers. In the means usually adopted, water is carried in specially made carts or vans, and is sprinkled on to the road through a perforated distributor in connection with each cart. Too much water should not be put on to the road at a time, or it will become sloppy and unsuited to traffic. There is practically no saving in time by putting a lot of water on to a road; the cart is more rapidly discharged, the water is wasted, and greater cost in labour entailed. Each gallon of water used should water four square yards of road.

Fig. 55 shows a patent four-valve box distributor manufactured by Messrs. Wm. Glover and Sons, Limited, Warwick. The distributor is divided into four compartments, each fed from the cart by a separate valve and connecting pipe; the upper compartments have small holes and the lower compartments larger holes. As any of these compartments may be used alone, the road can be sprinkled with either much or little water, and narrow strips as well as wide roads can be economically watered by this arrangement.

Fig. 56 illustrates a Willacy's patent rotary water sprinkler, detached from a watering cart. This provides a very efficient means of watering roads—the water is evenly distributed over a wide area, and is especially suited for wide roads, whilst by altering the gear the spread can be reduced to suit almost any desired width. The rotary machine may be adopted with considerable advantage on roads of average width. It has been ascertained, after repeated trial, that a satisfactory sprinkling can be made by the use of this machine at about half the cost of the same work performed by ordinary distributor carts.

Light four-wheel vans are rapidly superseding the two-wheel watering carts, as by their use the heavy load is drawn with less strain on the horse. The watering done by a four-wheel van is also more regularly spread than that by a two-

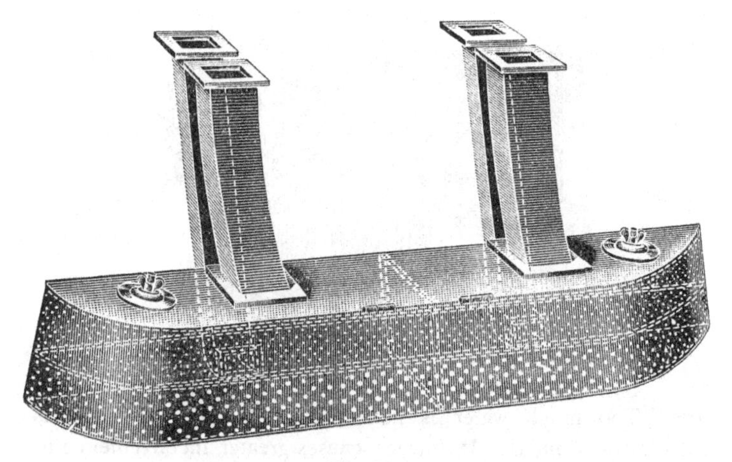

Fig. 55.

wheel cart. As the horse moves along it often produces an up-and-down movement in a two-wheel cart which causes the road to be watered in strips—strips alternately wet and dry—which gives the road an appearance of being out of repair. This does not occur in the use of a van. A great saving is effected in the use of a van over a two-wheel cart, by reason of the greater area of road that can be watered with each filling of the van. In hilly districts vans should not be made to carry more that 350 gallons; but where the roads are fairly level the capacity can be increased to advantage to 450 gallons. Two-wheel carts usually hold from 220 to 300 gallons. The cost of a 450-gallon van is about £60, of a 350-gallon van £46, and of a two-wheel cart £26. In cases where two-wheeled watering carts are used for hilly districts, carts with patent tipping gear are recommended; the advantage in their use is that, by means of a simple arrangement controlled by the driver, some of the dead weight of the load is removed from the horse's back whilst going down hill.

In Paris, and a few towns in this country, watering is frequently done by hose attached to the fire hydrants in the street. Metal pipes with flexible joints are generally adopted, or if hose pipe is used it should be protected by a coil of thick wire. Ordinary hose pipe, owing to the severe friction it receives, soon wears out. The hose method is not to be recommended in preference to

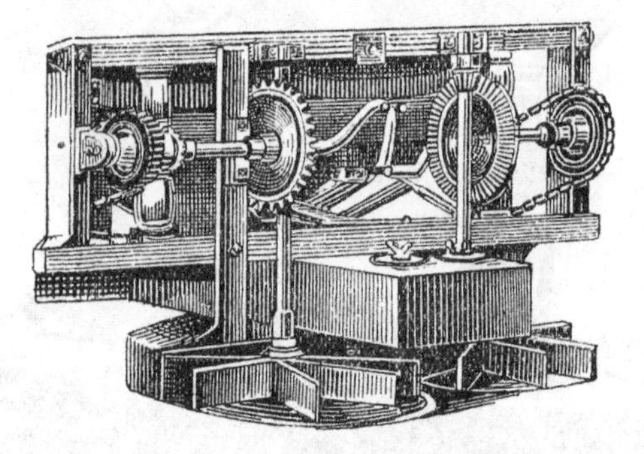

Fig. 56.

carts. Too much water is generally used, and as a rule is unevenly distributed. It, further, causes greater inconvenience to the traffic than watering carts, and the use of water from any other sources than that from the water mains is precluded.

The addition of a small quantity of soluble disinfectant to the water used for road sprinkling imparts an agreeable odour, besides being of value as a purifier.

The sanitary condition of all roads, especially those of macadam and wood, would be much benefited by the regular use of a small quantity of disinfectant, and the purity of the atmosphere of the streets would be improved, and the spread of infectious diseases retarded. The adoption of this course is not a very costly one. The author has used many cases of Hope's Pynerzone blocks for this purpose with advantage. One block, which costs 1s., will slowly dissolve in a watering cart, and will last for nearly a full day's sprinkling. One Pynerzone block will produce 5000 gallons of sanitary fluid suitable for washing down cab-stands and flushing drains. These blocks are made from an extract of Norwegian pine, eucalyptus, and thymol in a concentrated form.

Sea water for road watering has during recent years been largely employed in seaside towns. Salt water for this purpose has some advantages over fresh. There is a considerable saving in the cost of sprinkling by the use of salt water when it is near at hand as compared with that of fresh; one sprinkling of the former will lay the dust for a length of time that would almost require two or three sprinklings of the latter. A road, after being sprinkled once or twice with sea water, will remain free from dust for some time after the road is practically dry, as the deliquescent salts contained in the water form a hard crust, which in a measure preserves the surface.

There are weighty considerations in favour of salt water over that of fresh for road-watering, viz., the work is less costly, the traffic is not so much interrupted by the watering carts, and the road surface is kept in a much better state for the traffic. The author has frequently inquired of jewellers, ironmongers, and others having shops in streets regularly sprinkled with sea water whether they consider their goods in any way affected by the water, and has never found the slightest proof of its being so; but, on the contrary, tradespeople invariably express their approval of the system.

However, it cannot be denied that there is another side to the question of using sea water on roads, which, however suitable it may be in dry and hot weather, is held by many surveyors to have a deleterious after-effect on macadam during wet and wintry weather, it having a tendency to rot the same.

The author's father in 1890 prepared a report on the question of watering streets with salt water, in the course of which he quoted extracts from reports of seaside towns who had adopted its use. As there are adverse opinions on this question, it may be useful to give a few of these quotations *in toto* :—

Hastings and St. Leonards.—"The roads have been watered about eighteen years with sea water. We find it a great advantage. The sea water keeps the roads damp for a much longer period than fresh water, and it is a capital thing for flushing sewers, gullys, &c."

Portsmouth.—"About two years ago we introduced salt water for the purpose of watering the streets and the supply of our public baths, and are of opinion that where salt water can be pumped at a cost not greater than fresh water would entail, it is better to use salt water than fresh for the purpose of watering streets. One load of sea water will go as far as three loads of fresh."

Weymouth.—"The principal streets of this borough are watered with sea water, giving great satisfaction. The salt forms a crust on the surface, which keeps the dust from rising."

Harwich.—"We have used salt water for street watering and flushing sewers since 1879 with very satisfactory results."

Grimsby.—"With respect to the advantages of using salt water on the roads, I think I may safely say that my Committee are so pleased with the results that they do not intend to revert to the fresh water, even when the company are in a position to supply it."

Great Yarmouth.—"10,000,000 gallons of company's fresh water had been used for watering roads and flushing sewers, at an average cost of £552 per annum, and it was estimated that the supply necessary to keep the sewers properly flushed, including the watering of the roads, would involve an outlay of at least £700 per annum for the supply of 14,000,000 gallons. Works were put down for the supply of sea water, costing £4500. The repayment of principal and interest, with working expenses and depreciation, comes to £410 per annum. There is also a saving of £80 per annum in horse hire; so that our expenditure is £330 per annum as against £700, and instead of 14,000,000 we have 30,000,000." Since this period the sea-water service has been much extended at Yarmouth.

Birkenhead.—"Salt water pumped from the sea has been used in this borough, especially the central portion, for a great number of years for street-watering purposes. A spread of salt water on a macadam road is as effective in laying the dust as three spreads of fresh water, so that the tradespeople rather prefer sea water for this purpose."

Blackpool.—"We use salt water for watering the streets and flushing sewers; we use nothing else. Our streets are well watered with sea water, and practically no grievance arises therefrom."

Ryde.—"We use salt water for watering streets, and find it very advantageous. One load of salt water is equivalent to two loads of fresh. It makes a hard crust on the road."

Worthing.—We have watered a considerable portion of our town with sea water for many years. It keeps the roads damp at least twice as long as fresh water, and forms a kind of crust on the surface of the roads, which not only keeps them together, but also prevents the dust from moving for a considerable time after they are dry."

Plymouth.—"We use sea water for road-watering as much as possible; it answers perfectly for macadam roads."

Liverpool.—"Sea water possesses the advantage of remaining moist for a greater length of time, especially when the atmosphere is slightly damp."

New Shoreham.—"Salt water is used for the purpose of watering roads, and likewise flushing. The roads are more lasting and much harder than roads watered with fresh water, and take twice as long to dry if fresh water had been used."

Whitby.—"We use sea water for watering over the greater portion of the town, and find that it answers much better than fresh."

Tynemouth.—"We use sea water in this borough for watering the streets and flushing sewers. It hardens and binds the surface of the roads, and one watering with salt water would serve equally as well as two or three times watering with fresh water."

CHAPTER XXXI.

PRIVATE STREET REPAIRS.

THE control of private streets, *i.e.*, streets not repairable by the Local Authority as regards their state of repair, &c., is provided for in various ways by Acts of Parliament. The 150th Section of the Public Health Act, 1875, is that under which most surveyors have worked, but more recently an improved and new working has been passed, with the object of abolishing a multitude of difficulties that so frequently followed in the wake of the older Act. This Act, known as the Private Street Works Act, 1892, has been largely adopted, and although it necessitates very much more clerical work than any previous Act, the methods of procedure are in many ways preferable to those formerly prescribed.

The following are the principal contents of each of these Acts :—

SECTION 150, PUBLIC HEALTH ACT, 1875.

"Where any street within any urban district (not being a highway repairable by the inhabitants at large), or the carriageway, footway, or any other part of such street, is not sewered, levelled, paved, metalled, flagged, channelled, and made good, or is not lighted to the satisfaction of the Urban Authority, such Authority may, by notice addressed to the respective owners or occupiers of the premises fronting, adjoining, or abutting on such parts thereof, as may require to be sewered, levelled, paved, metalled, flagged, or channelled, or to be lighted, require them to sewer, level, pave, metal, flag, channel, or make good, or to provide proper means for lighting the same within a time to be specified in such notice.

"Before giving such notice the Urban Authority shall cause plans and sections of any structural works intended to be executed under this section, and an estimate of the probable cost thereof, to be made under the direction of their surveyor ; such plans and sections to be on a scale of not less than 1in. for 88ft. for a horizontal plan, and on a scale of not less than 1in. for 10ft. for a vertical section, and, in the case of a sewer, showing the depth of such sewers below the surface of the ground: such plans, sections, and estimate shall be deposited in the office of the Urban Authority, and shall be open at all reasonable hours for the inspection of all persons interested therein during the time specified in such notice; and a reference to such plans and sections in such notice shall be sufficient without requiring any copy of such plans and sections to be annexed to such notice.

"If such notice is not complied with, the Urban Authority may, if they think fit, execute the works mentioned or referred to therein; and may recover in a summary manner the expenses incurred by them in so doing from the owners in default, according to the frontage of their respective premises, and in such proportion as is settled by the surveyor of the Urban Authority, or (in case of dispute) by arbitration in manner provided by this Act; or the Urban Authority may by order declare the expenses so incurred to be private improvement expenses.

"The same proceedings may be taken, and the same powers may be exercised in respect of any street or road of which a part is or may be a public footpath or repairable by the inhabitants at large, as fully as if the whole of such street or road was a highway, not repairable by the inhabitants at large."

Private Street Works Act, 1892.

(55 and 56 Vict., Ch. 57.)

Before this Act can be put into force by any Urban Sanitary Authority, it is necessary that the Act should be adopted in the prescribed manner which is here given, together with the modes of procedure required for the proper operation of the Act.

The illustration—Fig. 57—shows a miniature plan and sections drawn as a guide to show the requirements of the Act.

(1) The adoption shall be by a resolution passed at a meeting of the Urban Authority; and one calendar month at least before such meeting special notice of the meeting, and of the intention to propose such resolution, shall be given to every member of the

Authority, and the notice shall be deemed to have been duly given to a member of it if it is either—

(*a*) Given in the mode in which notices to attend meetings of the Authority are usually given ; or

(*b*) Where there is no such mode, then signed by the Clerk of the Authoriry, and delivered to the member or left at his usual or last known place of abode in England, or forwarded by post in a prepaid registered letter, addressed to the member at his usual or last known place of abode in England.

(2) Such resolution shall be published by advertisement in some one or more newspapers circulating within the district of the

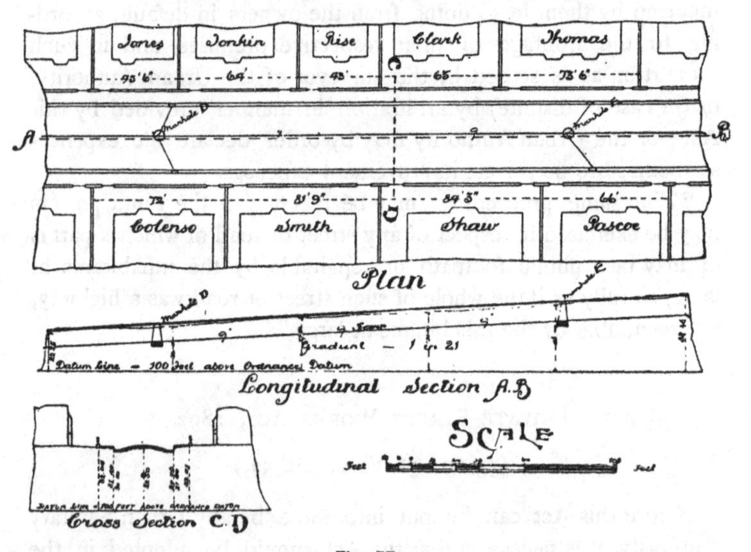

Fig. 57.

Authority, and by causing notice thereof to be affixed to the principal doors of every church and chapel in the place to which notices are usually fixed, and otherwise in such manner as the Authority think sufficient for giving notice thereof to all persons interested, and shall come into operation at such time not less than one month after the first publication of the advertisement of the resolution as the Authority may by the resolution fix, and upon its coming into operation this Act shall extend to that district.

(3) A copy of the resolution shall be sent to the Local Government Board.

(4) A copy of the advertisement shall be conclusive evidence of the resolution having been passed, unless the contrary be shown; and no objection to the effect of the resolution on the ground that notice of the intention to propose the same was not duly given, or on the ground that the resolution was not sufficiently published, shall be made after three months from the date of the first publication of the advertisement.

4. The Local Government Board may declare that the provisions contained in this Act shall be in force in any rural sanitary district, or any part thereof, and may invest a Rural Sanitary Authority with the powers, rights, duties, capacities, liabilities, and obligations which an Urban Authority may acquire by adoption of this Act, in like manner and subject to the same provisions as they are enabled to invest Rural Sanitary Authorities with the powers of Urban Sanitary Authorities under the provisions of Sec. 276 of the Public Health Act, 1875.

5. In this Act, if not inconsistent with the context—

The expression "Urban Authority" means an Urban Sanitary Authority under the Public Health Acts.

The expressions "urban sanitary district" and "rural sanitary district" mean respectively an urban sanitary district and a rural sanitary district under the Public Health Acts, and "district" means the district of an Urban Sanitary Authority or of a Rural Sanitary Authority, as the case may require.

The expressions "surveyor," "lands," "premises," "owner," "drain," "sewer," have respectively the same meaning as in the Public Health Acts.

The expression "street" means (unless the context otherwise requires) a street as defined by the Public Health Acts, and not being a highway reepairable by the inhabitants at large.

Words referring to "paving, metalling, and flagging" shall be construed as including macadamising, asphalting, gravelling, kerbing, and every method of making a carriageway or footway.

6.—(1) Where any street or part of a street is not sewered, levelled, paved, metalled, flagged, channelled, made good, and lighted to the satisfaction of the Urban Authority, the Urban Authority may from time to time resolve with respect to such street or part of a street to do any one or more of the following works (in this Act called private street works); that is to say, to sewer, level, pave, metal, flag, channel, or make good, or to provide proper means for lighting such street or part of a street; and the expenses incurred by the Urban Authority in executing private street works shall be apportioned (subject as in this Act mentioned) on the premises fronting, adjoining, or abutting on such street or part of a street. Any such resolution may include several streets or parts of streets, or may be limited to any part or parts of a street.

(2) The surveyor shall prepare, as respects each street or part of a street,—

(*a*) A specification of the private street works referred to in the resolution, with plans and sections (if applicable);

(*b*) An estimate of the probable expenses of the works;

(*c*) A provisional apportionment of the estimated expenses among the premises liable to be charged therewith under this Act.

Such specification, plans, sections, estimate, and provisional apportionment shall comprise the particulars prescribed in Part I. of the Schedule to this Act, and shall be submitted to the Urban Authority, who may by resolution approve the same respectively with or without modification or addition as they think fit.

(3) The resolution approving the specifications, plans, and sections (if any), estimates, and provisional apportionments, shall be published in the manner prescribed in Part II. of the Schedule of this Act, and copies thereof shall be served on the owners of the premises shown as liable to be charged in the provisional apportionment within seven days after the date of the first publication. During one month from the date of the first publication the approved specifications, plans, and sections (if any), estimates, and provisional apportionments (or copies thereof

certified by the surveyor), shall be kept deposited at the Urban Authority offices, and shall be open to inspection at all reasonable times.

7. During the said month any owner of any premises shown in a provisional apportionment as liable to be charged with any part of the expenses of executing the works may, by written notice served on the Urban Authority, object to the proposals of the Urban Authority on any of the following grounds; (that is to say),

(*a*) That an alleged street or part of a street is not or does not form part of a street within the meaning of this Act;

(*b*) That a street or part of a street is (in whole or in part) a highway repairable by the inhabitants at large;

(*c*) That there has been some material informality, defect, or error in or in respect of the resolution, notice, plans, sections, or estimate;

(*d*) That the proposed works are insufficient or unreasonable, or that the estimated expenses are excessive;

(*e*) That any premises ought to be excluded from or inserted in the provisional apportionment;

(*f*) That the provisional apportionment is incorrect in respect of some matter of fact to be specified in the objection or (where the provisional apportionment is made with regard to other considerations than frontage as hereinafter provided) in respect of the degree of benefit to be derived by any persons, or the amount or value of any work already done by the owner or occupier of any premises.

For the purposes of this Act joint tenants or tenants in common may object through one of their number authorised in writing under the hands of the majority of such joint tenants or tenants in common.

8.—(1) The Urban Authority at any time after the expiration of the said month may apply to a Court of summary jurisdiction to appoint a time for determining the matter of all objections made as in this Act mentioned, and shall publish a notice of the time and place appointed, and copies of such notice shall be served upon the objectors; and at the time and place so appointed any such Court may proceed to hear and determine the matter of all such objections in the same manner as nearly as may be, and with the same powers and subject to the same provisions with respect to stating a case, as if the Urban Authority were proceeding summarily against the objectors to enforce payment of a sum of money summarily recoverable. The Court may quash in whole or in part or may amend the resolution, plans, sections, estimates, and provisional apportionments, or any of them, on the application either of any objector or of the Urban Authority. The Court may also, if it thinks fit, adjourn the hearing and direct any further notices to be given.

(2) No objection which could be made under this Act shall be otherwise made or allowed in any Court proceeding or manner whatsoever.

(3) The costs of any proceedings before a Court of summary jurisdiction in relation to objections under this Act shall be in the discretion of the Court, and the Court, shall have power if it thinks fit, to direct that the whole or any part of such costs ordered to be paid by an objector or objectors shall be paid in the first instance by the Urban Authority, and charged as part of the expenses of the works on the premises of the objector or objectors in such proportions as may appear just.

9.—(1) The Urban Authority may include in any works to be done under this Act with respect to any street or part of a street any works which they think necessary for bringing the street or part of a street, as regards sewerage, drainage, level, or other matters, into conformity with any other streets (whether repairable or not by the inhabitants at large), including the provision of separate sewers for the reception of sewage and of surface water respectively.

(2) The Urban Authority in any estimate of the expenses of private street works may include a commission not exceeding five pounds per centum (in addition to the estimated actual cost) in respect of surveys, superintendence, and notices, and such commission when received shall be carried to the credit of the district fund.

10. In a provisional apportionment of expenses of private street works the apportionment of expenses against the premises fronting, adjoining, or abutting on the street or part of a street in respect of

which the expenses are to be incurred shall, unless the Urban Authority otherwise resolve, be apportioned according to the frontage of the respective premises; but the Urban Authority may, if they think just, resolve that in settling the apportionment regard shall be had to the following considerations; (that is to say)

(*a*) The greater or less degree of benefit to be derived by any premises from such works;

(*b*) The amount and value of any work already done by the owners or occupiers of any such premises.

They may also, if they think just, include any premises which do not front, adjoin, or abut on the street or part of a street, but access to which is obtained from the street through a court, passage, or otherwise, and which in their opinion will be benefited by the works, and may fix the sum or proportion to be charged against any such premises accordingly.

11. The Urban Authority may from time to time amend the specifications, plans, and sections (if any), estimates, and provisional apportionments for any private street works, but if the total amount of the estimate in respect of any street or part of a street is increased, such estimate and the provisional apportionment shall be published in the manner prescribed in Part II. of the schedule to this Act, and shall be open to inspection at the Urban Authority offices at all reasonable times, and copies thereof shall be served on the owners of the premises affected thereby; and objections may be made to the increase and apportionment, and if made shall be dealt with and determined in like manner as objections to the original estimate and apportionment.

12.—(1) When any private street works have been completed, and the expenses thereof ascertained, the surveyor shall make a final apportionment by dividing the expenses in the same proportions in which the estimated expenses were divided in the original or amended provisional apportionment (as the case may be), and such final apportionment shall be conclusive for all purposes; and notice of such final apportionment shall be served upon the owners of the premises affected thereby; and the sums apportioned thereby shall be recoverable in manner provided by this Act, or in the same manner as private improvement expenses are recoverable under the Public Health Act, 1875, including the power to declare any such expenses to be payable by instalments.

(2) Within one month after such notice the owner of any premises charged with any expenses under such apportionment may, by a written notice to the Urban Authority, object to such final apportionment on the following grounds, or any of them :—

(*a*) That the actual expenses have without sufficient reason exceeded the estimated expenses by more than fifteen per cent.

(*b*) That the final apportionment has not been made in accordance with this section.

(*c*) That there has been an unreasonable departure from the specifications, plans, and sections.

(3) Objections under this section shall be determined in the same manner as objections to the provisional apportionment.

13.—(1) Any premises included in the final apportionment, and all estates and interests from time to time therein, shall stand and remain charged (to the like extent and effect as under section two hundred and fifty-seven of the Public Health Act, 1875) with the sum finally apportioned on them, or if objection has been made against the final apportionment with the sum determined to be due as from the date of the final apportionment, with interest at the rate of four pounds per centum per annum, and the Urban Authority shall, for the recovery of such sum and interest, have all the same powers and remedies under the Conveyancing and Law of Property Act, 1881, and otherwise as if they were mortgagees having powers of sale and lease and of appointing a receiver.

(2) The Urban Authority shall keep a register of charges under this Act and of the payments made in satisfaction thereof, and the register shall be open to inspection to all persons at all reasonable times on payment of not exceeding one shilling in respect of each name or property searched for, and the Urban Authority shall furnish copies of any part of such register to any person applying for the same on payment of such reasonable sum as may be fixed by the Urban Authority.

14. The Urban Authority, if they think fit, may from time to time (in addition and without prejudice to any other remedy) recover summarily in a Court of summary jurisdiction, or as a simple contract debt by action in any Court of competent jurisdic-

tion, from the owner for the time being of any premises in respect of which any sum is due for expenses of private street works the whole or any portion of such sum, together with interest at a rate not exceeding four pounds per centum per annum, from the date of the final apportionment till payment thereof.

15. The Urban Authority, if they think fit, may at any time resolve to contribute the whole or a portion of the expenses of any private street works, and may pay the same out of the district fund or general district rate or other rate out of which the general expenses incurred under the Public Health Act, 1875, are payable.

16. The incumbent or minister or trustee of any church, chapel, or place appropriated to public religious worship, which is for the time being by law exempt from rates for the relief of the poor, shall not be liable to any expenses of private street works as the owner of such church, chapel, or place, or of any churchyard or burial ground attached thereto, nor shall any such expenses be deemed to be a charge on such church, chapel, or other place, or on such churchyard or burial ground, or to subject the same to distress, execution, or other legal process, but the proportion of expenses in respect of which an exemption is allowed under this section shall be borne and paid by the Urban Authority.

17. All owners of buildings or lands, being persons who under the Lands Clauses Acts are empowered to sell and convey or release lands, may charge such buildings or lands with such sum as may be necessary to defray the whole or any part of any expenses which the owners of or any persons in respect of such buildings or lands for the time being are liable to pay under this Act and the expenses of making such charge, and for securing the repayment of such sum with interest may mortgage such buildings or lands to any person advancing such sum, but so that the principal due on any such mortgage shall be repaid by equal yearly or half-yearly payments within twenty years.

18. The Urban Authority may from time to time, with the sanction of the Local Government Board, borrow, on the security of the district fund and general district rates or other rate out of which the general expenses incurred under the Public Health Act, 1875, are payable, monies for the purpose of temporarily

providing for expenses of private street works, and the powers of the Urban Authority to borrow under the Public Health Acts shall be available as if the execution of private street works under this Act were one of the purposes of the Public Health Act, 1875.

19. Whenever all or any of the private street works in this Act mentioned have been executed in a street or part of a street, and the Urban Authority are of opinion that such street or part of a street ought to become a highway repairable by the inhabitants at large, they may by notice to be fixed up in such street or part of a street declare the whole of such street or part of a street to be a highway repairable by the inhabitants at large, and thereupon such street or part of a street as defined in the notice shall become a highway repairable by the inhabitants at large.

20. If any street is now or shall hereafter be sewered, levelled, paved, metalled, flagged, channelled, and made good (all such works being done to the satisfaction of the Urban Authority), then, on the application in writing of the greater part in value of the owners of the houses and land in such street, the Urban Authority shall, within three months from the time of such application, by notice put up in such street, declare the same to be a highway repairable by the inhabitants at large, and thereupon such street shall become a highway repairable by the inhabitants at large.

21.—(1) The Urban Authority shall keep separate accounts of all moneys expended and recovered by them in the execution of the provisions of this Act relating to private street works.

(2) All moneys recovered by the Urban Authority under this Act in respect of street works shall be applied in repayment of moneys borrowed for the purpose of executing private street works, or if there is no such loan outstanding then in such manner as may be directed by the Local Government Board.

22. No railway or canal company shall be deemed to be an owner or occupier for the purposes of this Act in respect of any land of such company upon which any street shall wholly or partially front or abut, and which shall at the time of the laying out of such street be used by such company solely as a part of their line of railway, canal, or siding, station, towing path, or works, and shall have no direct communication with such street;

and the expenses incurred by the Urban Authority under the powers of this Act which, but for this provision, such company would be liable to pay, shall be repaid to the Urban Authority by the owners of the premises included in the apportionments, and in such proportion as shall be settled by the surveyor; and in the event of such company subsequently making a communication with such street they shall, notwithstanding such repayment as last aforesaid, pay to the Urban Authority the expenses which, but for the foregoing provision, such company would in the first instance have been liable to pay, and the Urban Authority shall divide among the owners for the time being included in the apportionment the amount so paid by such company to the Urban Authority, less the costs and expenses attendant upon such division, in such proportion as shall be settled by the surveyor, whose decision shall be final and conclusive. This section shall not apply to any street existing at the date of the adoption of this Act.

23. All expenses incurred or payable by an Urban Authority and a Rural Sanitary Authority respectively in the execution of this Act, and not otherwise provided for, may be charged and defrayed as part of the expenses incurred by them respectively in the execution of the Public Health Acts.

24. All powers given to a Local Authority under this Act shall be deemed to be in addition to and not in derogation of any other powers conferred upon such Local Authority by any Act of Parliament, law, or custom, and such other powers may be exercised in the same manner as if this Act had not been passed.

25. Neither Sections one hundred and fifty, one hundred and fifty-one, and one hundred and fifty-two of the Public Health Act, 1875, nor Section forty-one of the Public Health Acts Amendment Act, 1890, shall apply to any district or part of a district in which this Act is in force.

26. This Act shall not extend to prejudice or derogate from the estates, rights, and privileges of the Conservators of the River Thames, or render them liable to any charges or payments in respect of any of their works on or upon the shores of the River Thames.

THE SCHEDULE.

PRIVATE STREET WORKS.

PART I.

PARTICULARS TO BE STATED IN SPECIFICATIONS, PLANS AND SECTIONS, ESTIMATES, AND PROVISIONAL APPORTIONMENTS.

Specifications.—These shall describe generally the works and things to be done, and in the case of structural works shall specify as far as may be the foundation, form, material, and dimensions thereof.

Plans and Sections.—These shall show the constructive character of the works, and the connections (if any) with existing streets, sewers, or other works, and the lines and levels of the works, subject to such limits of deviation (if any) as shall be indicated on the plans and sections respectively.

Estimates.—These shall show the particulars of the probable cost of the whole works, including the commission provided for by this Act.

Provisional Apportionments.—These shall state the amounts charged on the respective premises and the names of the respective owners, or reputed owners, and shall also state whether the apportionment is made according to the frontage of the respective premises or not, and the measurements of the frontages, and the other considerations (if any) on which the apportionment is based.

PART II.

PUBLICATION OF NOTICE.

Any resolution, notice, or other document required by this Act to be published in the manner prescribed by this schedule shall be published once in each of two successive weeks in some local newspaper circulating within the district, and shall be publicly posted in or near the street to which it relates once at least in each of three successive weeks.

CHAPTER XXXI.

THE following are specimen reports to be adopted by the Surveyor under the Private Street Works Act, 1892, commencing with the necessary preliminary report and terminating with the final apportionment:—

Surveyor's first report addressed to the Urban Sanitary Authority.

PRIVATE STREET WORKS ACT, 1892.

I beg to report that the following named street, viz.,
is not properly sewered, levelled, paved, metalled, and flagged, and I consider that this is a proper case in which the Private Street Works Act, 1892, should be applied.

The Urban Sanitary Authority, having considered the report, will pass a resolution pursuant with the Private Street Works Act, 1892, that the Surveyor do prepare and submit to the Urban Sanitary Authority—

(*a*) "A specification of the Private Street Works referred to in the resolution with plans and sections (if applicable).

(*b*) "An estimate of the probable expenses of the works.

(*c*) "A provisional apportionment of the estimated expenses among the premises liable to be charged therewith under the Private Street Works Act, 1892.

"Such specifications, plans, sections, estimates, and provisional apportionment to comprise the particulars prescribed in Part I. of the Schedule to this Act."

On completion of the necessary plans, &c., the Surveyor should report to the Urban Sanitary Authority, using the wording of the foregoing resolution, and continuing: And that the following operations, hereafter specified, are necessary for executing such work under the provisions of the Private Street Works Act, 1892, and I herewith submit specification, plans, sections, estimate, and provisional apportionment for the same.

The Urban Sanitary Authority, upon receiving the report of the Surveyor, accompanied with the necessary specification, plans, sections, estimate, and provisional apportionment, prepared in accordance with their former resolution, approve or otherwise of the specification, plans, sections, estimate, and provisional apportionment, prepared and submitted by the Surveyor.

This resolution is then published in a local newspaper, and publicly posted in or near the street to which it relates, according to Part II. of the schedule of the Act. This is in order to enable owners to give notice of objection before the work is executed, which is a much better way than that of objecting after the work is completed.

The Surveyor's next report to the Urban Sanitary Authority should be as follows :—

PRIVATE STREET WORKS ACT, 1892.

I beg to report that the plans, specification, estimate, and provisional apportionment for putting road into repair, having been approved of by the Urban Sanitary Authority, and the resolution approving same published in the prescribed manner, and copies served on the owners of the properties affected, the period of which having expired, I shall be glad of your further instructions in the business.

The Urban Sanitary Authority here give instructions for the work to be proceeded with.

Finally, on the completion of the work, the Surveyor should report as follows :—

PRIVATE STREET WORKS ACT, 1892.

I beg to submit the final apportionment for the following named road, viz., which has been satisfactorily put in repair under the powers of the Private Street Works Act, 1892.

CHAPTER XXXII.

SPECIFICATION CLAUSES.—ROADS AND PATHS.

Clauses Suitable for Roads for Heavy Traffic.

Contractor to perform all the necessary excavations and filling as may be required to bring the surface of the street to its proper form and level, as shown in the sections, and to cart away to a shoot to be provided by the contractor, or (to such place as directed by the engineer) any surplus material that shall not be required for filling-in purposes.

Construct the street to the widths shown on the plans.

Contractor to provide and fix kerbing and channelling; the kerbing to be best selected 12in. by 8in. granite kerb in lengths of not less than 2ft. 6in. laid in true lines on its natural bed upon 4in. thickness of 1 to 6 of Portland cement concrete. The kerb to be square dressed on top, and 4in. on back, and tooled with a ¼in. batter on breast, the end joints to be squared 6in. down. The joints to be made and pointed with good hydraulic lime mortar.

The kerb when laid shall be level with the finished surface of the footpath, and shall stand 4in. higher than the surface of the channelling.

The channelling to be of granite stone 12in. wide, not less than 4in. thick, and 3ft. in length, the same to be hammer dressed on surface with squared ends and sides, and to form close joint with the kerb and each other.

The channelling to be laid on a thickness of 4in. of 1 to 6 Portland cement concrete, and to the camber of the road, and the joints made throughout with 1 to 2 Portland cement mortar.

At the street corners the kerbs and channelling are to be brought to a true radius to approved template.

Contractor to provide proper gully catch pits at the positions indicated on the plans, the same to be Sykes' Patent "Granitic stoneware" of 60 gallons capacity. A collaring of brickwork in cement based on a 9in. layer of cement concrete surrounding the gully is to be placed over the gully to carry a Sykes' patent grating 12in. by 15in. Each gully to be connected with the sewers by means of 6in. socketed salt-glazed stoneware pipes jointed in cement.

Well roll the road bed with a 10-ton steam road roller, and upon this formation close pave the full width of the carriageway with good hard paving stones, the same not to be less than six inches across in any direction. The stones are to be set on a flat side by hand, and to be solidly rammed and bound together by the aid of scabblings; the pitching to be laid with a surface camber from kerb to kerb similar to the curvature required for the finished road.

The foundation to be covered with a 9in. thickness of good durable metalling, broken to pass all ways through a 2in. ring, and any excess of small stuff will be rejected by the engineer. The 9in. covering to be put on and rolled in two layers, the first layer of 5in. to be thoroughly consolidated by rolling with a 12-ton steam roller before the second layer is applied, the remaining thickness of 4in. is then to be laid, and with the addition of a small quantity of binding material rolled as before to a finished surface.

Contractor to properly form the bed of footways, the same to be made up with hard core and rammed to a solid and even surface, over which a 1in. layer of sand is to be placed graded to the fall of the finished footpath. The footpath is to be of Victoria stone 2in. in thickness, bedded on the layer of sand and set perfectly flush with bonded joints, and with a fall towards the kerb of 1in. in 4ft. The joints on completion are to be grouted in hydraulic lime grout and pointed in mortar.

Clauses suitable for Roads of Medium or Light Traffic.

Excavate the ground on the site of the street to the depths shown on the sections, and cart away all surplus stuff.

Correctly form the foundations of carriageways and footways, according to the sections, and well roll the surface as formed with a ten-ton steam roller.

Provide and lay water channelling 12in. in width, executing same in two courses of 4in. by 6ft. granite pitchers on 4in. of Portland cement concrete. The pitchers to be laid to line with the road, and to be jointed in Portland cement.

Provide and fix squared and tooled Norway granite kerb 12in. by 8in. laid flat, on 4in. of Portland cement concrete, and joint same in cement. No kerb stones to be less than 2ft. in length.

(Or) ovide and lay concrete channelling 16in. in width and 6in. in depth, and kerb to match 12in. deep and 6in. wide. The channel and kerb to consist of Portland cement, two parts of clean, sharp sand, and three parts of clean stone broken to an inch gauge. The exposed surfaces of both channel and kerb to consist of 1in. rendering of cement mortar, mixed in proportion of 1 of Portland cement to 1 of sand, the same being applied whilst the concrete is in a plastic state. The contractor to mould the channel and kerb into shape to conform on completion with the drawings. The channel and kerb are to have joints at intervals of every 10ft. to allow for expansion and contraction.

Construct the gullies for surface water where shown on plans in best stock brickwork in cement, with cast iron frames and gratings, and properly connect same with the sewers by means of glazed and socketed stoneware pipes 6in. in diameter jointed in gaskin and Portland cement. Provide and fix at outlet from each gully an approved stoneware gully block to trap the gully from the sewer. Well render the interior of the gullies in Portland cement . in thickness.

Line the carriageways to a thickness of 6in. with best road metalling broken to a 2in. gauge, and well roll same with a steam road roller in two layers, with a small quantity of binding material.

Construct the footways as shown with proper tar concrete, well rolled in layers, 2¼in. thickness, of coarse stuff and 1in. of finings, well rolled down to 2¾in. in thickness when finished.

CLAUSES SUITABLE FOR WOOD-PAVED ROADS.

Contractor to excavate the ground on the site of the work to the required level, the excavated road material consisting of macadam, together with all channelling, &c., to remain the property of the Council, and to be carted to a site to be provided by the engineer, but all surplus material not required by the engineer shall be removed elsewhere by the contractor.

Contractor to give proper and sufficient notice to the gas, water, electric, and other companies, to enable them to attend to their mains and services before the foundation works are proceeded with.

Contractor to lower any gas, water, or other pipes, that may be exposed in carrying out the work, so that they may be left below the concrete foundation. He shall also alter any manhole stop-taps, or valve covers or gullies, to the level of the wood paving.

Contractor to thoroughly roll the ground with a heavy roller, and fill up all soft places with hard core well rolled in. He shall then construct a foundation of Portland cement concrete to a thickness of not less than 6in. The concrete to be composed of one part of cement to two parts of fine sharp and clean sand, and three of clean stone broken to a 1½in. gauge, which shall be thoroughly incorporated on a platform whilst dry and again whilst wet. The concrete is to be flushed up to an even surface, of a contour similar to that required for the carriageway when finished.

The wood blocks to be first quality Jarrah (or Karri), well seasoned and free from sap, cracks, and shakes, and to be cross-cut and accurately spaced 3in. wide, 4½in. deep, and 9in. long.

The channel courses on either side of the road to be of the same paving blocks, and to be laid with a slip of deal ¾in. thick between each course, and after the transverse paving is laid these laths are to be removed and the spaces next the kerbs thoroughly filled with a mixture of tar pitch and sand, and the remaining joints filled with clay.

The wood paving proper to be laid with the blocks set transversely in true lines and to proper camber, and each block as laid is to be dipped to half its depth into a hot composition of tar pitch and creosote oil of proper consistency.

The paving, with exception of the channel courses, to have close joints, the blocks being driven together as laid with a mallet, and before completion the whole of the joints are to be grouted with Portland cement of the proportion of 1 of Portland cement and 1 of sand. The surface when finished to be smooth and even, and to be top dressed with fine gravel or grit.

CLAUSES SUITABLE FOR COMPRESSED ASPHALT PAVING.

Excavate the ground the full width of carriageways and proper depth, and to similar camber as finished surface of paving, the excavated macadam to be removed to a site to be provided by the engineer, and to be the property of the Council, and the other excavated material to be removed from the site and disposed of by the contractor.

Contractor to give sufficient notice to the gas, water, electric, and other companies to enable them to attend to their mains and services before the foundation works are proceeded with, and the contractor shall see that all such pipes and services above the excavated bed are lowered, so that they may be left below the concrete foundation.

Contractor to regulate all manhole and other covers to the level of the finished pavement.

Contractor to roll the foundation bed until it is perfectly firm and unyielding, filling in all weak spots with hard core.

On the bed thus prepared place a layer of 6in. deep of Portland cement concrete, composed of four parts of clean sharp stone, broken to a 2in. gauge, two parts of clean, sharp sand, and one part of approved Portland cement, carefully united on a proper mixing platform, twice whilst dry and twice whilst wet, the water being applied to the making through a rose. The surface to be of a similar camber to finished pavement and "staffed" to templates. The face of the concrete to be floated whilst in a plastic state with a 1in. coat of Portland cement and sand, gauged in equal proportions.

Upon the concrete foundation becoming quite dry, spread an even layer of 2½in. thickness of heated natural rock asphalte powder over the surface, and well ram same with heated iron punners until thoroughly compressed, the surface shall then be smoothed with proper hot iron tools, and finally rolled till quite cool. Previous to the traffic being admitted on to the pavement, the contractor shall cover it with a thin layer of sand.

FOOTPATHS. CONCRETE LAID *in situ.*

Contractor to excavate the present footpath to the depth shown on the sections and remove all surplus materials from the site.

The surface shall be regulated in accordance with the longitudinal and cross sections, and well rammed. The prepared bed to be covered with a 1in. layer of clean, sharp sand, on which construct Portland cement paving, 3in. thick, in the following manner: A layer of 2in. thick, to be composed of 3 parts clean gravel that will pass through a sieve with ½in. mesh, 2 parts clean sharp sand, and 1 part of approved Portland cement, carefully measured and thoroughly mixed on a suitable platform, twice in a dry state and twice in a wet state; and the paving to be finished by adding, whilst the bottom is still wet, 1in. of topping composed of clean sharp sand, thoroughly washed, and Portland cement, well mixed in the proportion of 1 to 1. The face to be flouted and finished in approved manner. The concrete paving is to have a fall towards the kerb of ¼in. to the foot. The paving is to be constructed *in situ* in sections, with wood divisions between each section. These sections are to be laid alternately, and allowed to harden before laying the ones adjoining, when the wood screeds shall be removed, and strips of brown paper used in their stead and left there permanently. The blocks to be 4ft. by 3ft., and laid so as to equally break joint. The face of the paving to be protected whilst laying against the effect of the weather or sun.

SPECIFICATION CLAUSES FOR QUARRYING OF ROAD STONE.

The contractor shall follow on the course of rock known as basalt, in such manner and at such levels as directed by the engineer, and he shall raise a quantity not less than tons per week during the months of unless otherwise directed; the total quantity required is approximately per annum, but no guarantee is given as to quantity. The contractor shall break the rock when quarried to a size not to exceed 8in. when measured either way.

The rock is to be blasted to as small an extent as possible, in order that the material shall not be unduly shaken in the process. Machine drills shall be used and sufficient drill holes shall be arranged so as to give the best results with each charge of dynamite or nitroglycerine. The holes, prior to firing, shall be covered with tree branches, and if necessary weighted by old chains to prevent stones flying; and should any damage to life or property occur in blasting, either through carelessness, neglect, or any other cause, the contractor shall be liable for all damages, and shall compensate the engineer to the full amount of any charges or expenses he may have incurred by reason of such accidents. Such amount shall be deducted from any moneys that may be or become due to the contractor, or shall be paid by the contractor, or failing these it shall be recoverable from him.

The contractor shall remove all inert earth and overburden immediately over where quarrying operations are proceeding, and shall cart the same from the site of the quarry.

The contractor shall set back the fence round the top of quarry whenever it is necessary, and such fence is not at any time to be less than 6ft. from the edge of the rock. The contractor shall be equally responsible for any accidents that may occur through the improper fixing of the fencing, as for carelessness in blasting operations.

The contractor shall provide all tools, implements, planks, blasting materials, team and manual labour that may be necessary for the proper working of the quarry under the instructions and to the satisfaction of the engineer.

The contractor to state his price per ton, which shall include all the works mentioned in this specification.

The contractor to give a surety of £25 for the due performance of his contract. Payments will be made at the rate of 85 per cent. of the work done, but no payment will be allowed on any quantity less than 100 tons.

CONTRACT FOR ROAD STONE.

Tenders required for   yards of broken road stone, to be delivered within the period of three calendar months from the date of contract, the tenders to state from where the stone is obtained.

The material to be delivered and squared up by the contractor in heaps where directed, and measured by the engineer, whose measurement shall be accepted by the contractor as final; no quantity less than 100 yards will be measured.

The stone shall be carefully blasted for road purposes, and broken by hand (machine) to pass in all directions through a 2in.

ring, and if not so broken the engineer shall be at liberty to reject the bulk, or may employ men to break the stones to the stipulated gauge, and deduct the cost of same from money due to the contractor.

A sample quantity of each description of stone to be delivered shall be sent to the Engineer at the same time as the tender.

A penalty of will be inflicted on the contractor for breach of contract, and a sum of per day will be deducted for each and every day during which the contract remains incomplete after the expiration of three calendar months from the date of the annexed contract.

Sealed tenders endorsed "Tender for Road Stone," &c. &c.

The lowest or any tender will not necessarily be accepted.

CHAPTER XXXIII.

SPECIFICATION CLAUSES.—SEWERS AND SEA WALLS.

CLAUSES SUITABLE FOR STONEWARE, IRON AND BRICK SEWERS IN NEW ROADS, MANHOLES, GULLIES, &C.

The contractor shall supply any earthenware sewer pipes that may be required; such pipes shall be tubular glazed stoneware socket pipes manufactured by Messrs. Henry Doulton and Co., the Albion Clay Company, Limited, Woodville, or other make, to be approved by the engineer. The stoneware pipes shall have the thickness and dimensions when burnt as follows:—

Sizes of Pipes.		Thickness of Material.		Depth of Socket.
6in.	...	$\frac{3}{8}$in.	...	1$\frac{3}{4}$in.
8in.	...	$\frac{4}{8}$in.	...	1$\frac{7}{8}$in.
9in.	...	$\frac{9}{16}$in.	...	2in.
&c.	...	&c.	...	&c.

All pipes shall be perfectly straight and truly cylindrical, salt glazed inside and outside, free from cracks and flaws, and perfectly burnt. Any pipes that are not perfectly straight and truly cylindrical, well and uniformally glazed, free from cracks and flaws, and perfectly burnt, will be rejected, and the contractors shall supply other pipes in their place. The contractor shall place in the sewers as the work proceeds proper junction pipes where shown on the plans, and those which are not required for immediate use shall be stopped with a disc of earthenware cemented into the socket.

The contractor shall supply bricks of equal quality to the sample brick deposited in the office of the engineer for the guidance of the contractor; the bricks for the circular manholes shall be radiating bricks of similar quality to the deposited brick. The bricks shall be hard burnt, of true shape, and when laid in the work no joint in the brickwork shall exceed $\frac{5}{16}$in. in thickness.

The contractor shall construct the brick sewers shown on plans in brickwork and concrete work, neatly turned on proper centres, and the inverts shall be constructed of approved invert sewer blocks bedded in concrete. The contractor shall take special care in executing the concrete portion of any sewer, or other work, in order to ensure the uniform setting of the concrete before the ground is filled in.

The contractor shall provide the cast iron pipes where shown on the plans. All pipes shall be truly cylindrical, and the spigot end of every pipe shall fit into the socket, leaving only the thickness of joint specified. The spigot end of each pipe shall be furnished with a projecting fillet, which shall fit into the socket, having not more than $\frac{1}{16}$in. clearance all round the pipe. Any special pipes that may be required shall be truly shaped, and care shall be exercised to see that they joint properly with the straight pipes. The sectional area of every pipe shall be truly concentric. The pipes shall be cast vertically with the socket downwards, and without the use of core nail. The metal used shall be free from cinder or other inferior iron, and shall be strong, tough, and close grained, and the pipes shall be free from scoria, air holes, or other imperfections of casting.

The iron pipes shall be well cleaned and coated whilst hot with Dr. Angus Smith's solution.

The pipes shall be as nearly as possible of the weight set forth in the schedule of particulars, and any pipe deviating more than the specified number of pounds shall be rejected.

The cement used in the works shall be fresh-burnt, but not hot, finely-ground Portland cement, weighing not less than 116 lb. per striked bushel, when filled from a hopper having a fall of not more than 1ft., and shall be capable—when mixed with water, and set in a proper mould, and after seven days' immersion in water—of bearing a tensile strain of 350 lb. to the square inch. It shall be ground to pass through a sieve having 2500 meshes to the square inch, and not leave more than 10 per cent. residuum in the sieve.

The sand used shall be clean, sharp, and silicious sand, free from all organic or saline matters.

The concrete used upon the works shall be properly gauged, the stone, sand, and cement being separately measured in the proportion of 3 parts broken stone, 2 parts sand, and 1 part cement. The stones for the concrete shall be broken to pass through a sieve with 1½in. meshes, and freed from all sand and washed if requisite. The aggregate shall be brought together on a platform, and carefully mixed and turned twice whilst dry and twice whilst wet.

The mortar used in the construction of the sewers, manholes, ventilators, and gullies shall be composed of 2 parts of clean sharp sand, and 1 part of best Portland cement, thoroughly washed with clean water. The mortar is to be conveyed fresh to the works as required for use.

The contractor shall excavate the ground for the construction of the sewers, manholes, and gullies, to the depths and inclinations shown in the sections. The floor of every trench shall have a true grade throughout, and shall be made in straight lines as shown on the plans. All soft places in the bottom of trenches shall be removed by excavating down to a solid foundation, and which shall be made up to the proper level with concrete or hard core as directed. The floor of the trenches shall be formed into recesses cut at intervals for the reception of the sockets of the pipes. In cases where brick sewers are to be constructed the floor of the trench shall be formed to the proper radius of the sewer.

In all cases the trenches for the sewers shall be open cut.

(Or) the trenches on the line of the road, where under 14ft. deep, shall be open cut.

(Or) the trenches through fields may be alternate open and adit work, but no sewer shall be laid in a heading until communication is made between one open trench and another.

All adits shall be well filled on completion of the sewer, as the work proceeds.

The stoneware pipe sewers shall be jointed by first forcing two strands of tarred spun gaskin into the joints, the rings to be sufficiently thick to tightly fit the annular space between the sockets and spigots. The annular space shall then be solidly filled with neat Portland cement (or cement and sand mixed in equal proportions answers very well), which shall be forced into the socket so as to fill it, and a fillet of cement shall then be worked round the outside of the joint. The under section of the joint shall always be first filled.

In cases where the sewer is in proximity with a water-supply well, the following clause should be inserted :—

Every joint of the stoneware pipes shall be further protected by placing on the outer side of the cement joint well-tempered and tenacious clay, so as to completely surround the joint, and for this purpose not less than the following quantities of clay shall be used round every joint, viz.:—6in. joint, ½ cubic foot; 8in., ¾ cubic foot; 9in., 1 cubic foot, &c. The contractor shall guarantee the water-tightness of all the pipe sewers.

The contractor shall lay the pipes in perfectly true lines, with a regular fall trom point to point of the same.

The contractor shall joint all iron sewer pipes by first driving cold rings of drawn lead into the sockets of each pipe until it is half filled, after which the remaining space shall be run in with melted lead, and the whole finally caulked up with a caulking tool, and on completion every joint shall be full of lead.

Before the trenches are re-filled, the contractor shall place the finest selected material under and immediately over and around every pipe or sewer, care being taken that the under filling is perfectly firm before any further filling is done. No lumps or stones shall be put round the sewer or be thrown into the trenches, until the pipes have been protected by the finer filling.

The utmost care shall be taken in all respects not to disturb, break, or damage the jointed sewers.

The contractor shall refill the trenches with the material taken therefrom, and shall be well rammed in layers until it is com pletely consolidated. No soft material shall be placed within 2ft. of the surface of any roadway, but the harder material shall be put on one side for filling the top of all trenches, manholes, &c. Contractor shall provide a suitable site for the disposal of all surplus earth, and shall remove all such earth immediately after every trench is filled up.

The contractor shall build the manholes where shown on the plans in accordance with the detailed drawings, the floor of all manholes shall be constructed of brickwork in cement laid on 12in. of Portland cement concrete. Proper channels shall be formed across the manholes in cement, and the benchings and walls to a height of 6ft. shall be rendered in cement 1in. in thickness. The contractor shall provide and build in all foot irons for the manholes, these shall be of cast iron built at least 9in. into the walls.

The contractor shall provide and fix all the manhole covers as shown on the plans, 20in. clear diameter, having circular covers filled with elm blocks on surface.

The contractor to provide and fix Crosta's patent iron surface water gullies where shown on plans of 57 gallons capacity, with cast iron gratings complete, and connect same with the sewers by means of 6in. stoneware drain pipes jointed in cement.

SPECIFICATION CLAUSES FOR SEA WALLS.

The contractor shall dig the excavations as shown on the plans in sections, so that each section may be completed in one tide, and the foundations shall be rendered dry for the reception of the concrete and masonry, which are to be filled in and left well protected on the same tide. All foundations shall be inspected by the Engineer and approved of by him.

The cement used in the works shall be fresh-burnt, finely-ground Portland cement weighing not less than 116 lb. per striked bushel when filled from a hopper having a fall of not more than 1ft. and shall be capable—when mixed with water and set in a mould, and after seven days' immersion in water—of bearing a tensile strain of 350 lb. to the square inch when tested on a section of 2¼ square inches area. It shall be ground sufficiently fine to pass through a sieve having 2500 meshes to the square inch, and not leave more than 10 per cent. residuum in the sieve. The contractor shall provide all requisite labour and materials for testing the cement. Any matter in dispute in reference to the quality of the cement may be referred by the Engineer to some expert, and determined by him, and the cost of such reference shall be paid by the contractor.

All sand used in mixing with the cement shall be clean, sharp, and silicious, free from organic matter, and it shall be washed by the contractor before it is used, should the engineer order the same to be done.

All concrete used upon the works shall be properly gauged; the stone, sand, and cement shall be separately measured. The stones for the concrete shall be approved by the engineer, and before use shall be broken to pass through a sieve with 1½in. meshes. All the stones shall be sifted free from sand and washed if requisite. The aggregate shall be brought together on a platform and mixed in the proportion of three measures of stone, two of sand, and one of cement; these materials shall be well incorporated on the platform provided, before mixing with water, and clean water shall be used for mixing, as little water as practicable being used, but sufficient to make the materials plastic. The concrete in all cases shall be carefully lowered or placed into position, and it shall not be shot into the the work from any elevation, and when in the work the contractor shall well compact the same so as to ensure its being solid and uniform throughout.

CONCRETE-FACED WALL.

A special concrete shall be provided for facing the work namely, mixed with two measures of stone, one of sand, and one of cement. The concrete shall be carefully placed in position against the framed mould to the full height from the level of the rock foundation and to a thickness of 1ft. throughout.

The wall shall be formed to the section and profile as shown on the plans, and shall be constructed in bulk. Provide and fix suitable timber frames and boards to contain and secure the concrete to its proper section while setting; the frame to be so constructed as to leave the face of the wall smooth when finished.

The contractor shall construct a porous drain in the formation behind the wall and as near to the base as possible, for the purpose of passing underground water; this drain shall be carried through the wall at its lowest part by means of a wrought iron pipe, the end of which shall be protected by a lead-faced balance valve for the purpose of keeping out the sea.

The contractor shall carefully make good on the top of the wall with fine concrete trowelled to a smooth surface, whilst a large fillet shall be formed outside in accordance with the drawing.

GRANITE OR STONE-FACED WALL.

The contractor shall face the wall with granite ashlar in level courses, alternate header and stretcher, with scappled beds and joints, and regular rough-picked faces.

The foundation shall consist of rock-faced granite, with good scappled beds and joints in courses, and no header in these courses shall be less than 2ft. in depth, and no stretcher shall be less than 1ft. 6in. in depth. These courses shall be set in Portland cement mortar gauged in equal proportions. Each course of the foundation work shall be carefully and solidly backed with concrete as it is laid to the width shown on the plans.

The profile of the wall shall be according to that shown in the drawings, and all stones shall be worked in courses to template. The facing stones of wall shall be carried out with alternate header and stretcher, and there shall be a bond of at least 6in. in the vertical joints of all courses. No course shall be less than 9in. in height, and no stretcher shall have a less bed than the height of the course to which it belongs. The bed and vertical joints shall be set in Portland cement mortar gauged in the proportion of one of Portland cement to two of clean sharp sand, and each course as it is laid shall be backed with concrete to the thickness shown on plans, and the contractor shall take great care that the concrete is well packed against the back of the wall, and that the heading stones are completely surrounded with that material. The face of the wall shall be pointed in neat cement as the work proceeds, each joint being sufficiently raked out to give a good key for the pointing.

No stone immediately under the coping stones shall have a less height than 15in., nor less bed than 18in.

The coping stones shall be of sound granite, with bevelled nose, as shown in drawings. The stones shall be 2ft. in height and 3ft. wide, and no stone shall be less than 3ft. in length. The top surface of coping shall be slightly splayed towards the sea. The vertical joints of all coping stones shall be grooved with a dovetailed chase, 6in. deep, each end of the stone, and at points corresponding with these grooves 1½in. holes shall be bored into the course immediately under the coping stones to a depth of 8in., into which 1in. iron bolts 18in. in length shall be inserted head downwards, and the space round the bolt shall be filled in with neat cement grout.

Upon the coping stones being set in position, and the joints effectually made, the opening formed by the grooves cut in the vertical joints of each stone, and which contains a length of the 1in. bolts previously fixed, shall be completely filled in with neat cement grout.

The contractor shall construct weep holes through the wall at its base, 30ft. apart, which shall consist of a wrought iron pipe carried through the full thickness of the wall, provided with a 4in. metal-faced, chained valve against the wall to keep out the sea.

APPENDIX I.

APPROXIMATE PRICES

OF

ROAD-MAKING MATERIALS.

CEMENT.

PORTLAND CEMENT.

A "central" of cement weighs 100 lb a cask contains four "centrals."

	£	s.	d.
Cost per central, d/d... ...	0	2	0
,, ton, d/d	1	15	0
,, cask, d/d	0	10	0

ROMAN CEMENT.

Cost per bushel, d/d	0	1	9
,, ton, d/d	1	14	6
,, cask, d/d	0	9	6

MEDINA CEMENT.

Cost per bushel, d/d	0	2	3

PARIAN CEMENT.

Cost per ton, d/d	3	10	0

KEEN'S CEMENT.

Cost per ton, d/d	3	12	0

MORTAR.

Per yard	0	15	0
Cement rendering suitable for footpaths, 3 to 1, 2in. thick, per yard super. ...	0	3	2

CONCRETE, per cubic yard.

	£	s.	d.
1 of lias lime to 6 of gravel and sand	0	7	6
If made with Thames ballast	0	10	6
1 of Portland cement to 6 of gravel and sand... ...	0	11	0
If aggregate consists of Thames ballast	0	14	6

CEMENT FOUNDATION FOR ROADS, FOOTPATHS, &c., LAID, INCLUDING EXCAVATION.

1 of Portland cement to 4 of gravel and sand, 6in. thick, 4s. per yard super.

Ditto, using coke breeze instead of gravel, 3s. 3d. per yard super.

Ditto, for finished footpaths, foundation 4in. thick, divided into sections, floated face, 1in. thick, using gravel, 4s. 6d. per yard super.

Using coke breeze and sand rendering, 4s. per yard super.

Extra for rolled vermiculated surface, 8d. per yard super.

Vermiculating Rollers.

Gilchrist's, 7in. long by 4in. diameter, 30s. each.

Brass jointing rollers, from 5s. to 9s. each.

ARTIFICIAL STONE FLAGGING.

Patent Victoria Stone.

	£	s.	d.
At the works, flagging 2in. thick, per foot super. ...	0	0	7

Stuart's Patent Granolithic Stone Flags.

	£	s.	d.
2in. thick, per foot super....	0	0	6
Laid *in situ* complete, per yard super.	0	4	6

Imperial Stone.

	£	s.	d.
Per yard super, laid	0	5	9

Adamant Paving Stone.

	£	s.	d.
Per yard super, laid	0	5	6

ASPHALTE.

Laid in or near London, but not including any foundation work.

Val de Travers Mastic Asphalte Paving.

	£	s.	d.
½in. thick, per yard super...	0	4	0
¾in. „ „ ...	0	5	0
1in. „ „ ...	0	6	9
2in. „ „ ...	0	12	6

Compressed Asphalte.

	£	s.	d.
¾in. thick, per yard super...	0	4	0
1in. „ „ ...	0	5	6
1¼in. „ „ ...	0	7	0
1½in. „ „ ...	0	8	0
2in. „ „ ...	0	10	0

Limmer Asphalte Paving.

Mineral rock asphalte.

	£	s.	d.
½in. thick, per yard super...	0	4	0
¾in. „ „ ...	0	5	0
1in. „ „ ...	0	6	6
1¼in. „ „ ...	0	8	0
1½in. „ „ ...	0	9	6
2in. „ „ ...	0	12	0

Asphaltic Limestone Concrete Paving.

	£	s.	d.
2½in. thick, per yard super.	0	2	4
4in. „ „ ...	0	3	4

Tar Paving.

	£	s.	d.
2½in. thick, per yard super.	0	2	3
3in. „ „ ...	0	2	6
3½in. „ „ ...	0	3	0
4in. „ „ ...	0	3	6
6in. „ „ ...	0	4	3
9in. „ „ ...	0	5	3

BRICK PAVING.

Number of bricks required per yard super:—

36 paving bricks laid flat.
82 „ on edge.
36 stock bricks laid flat.
52 „ on edge.
70 clinkers laid flat.
140 „ on edge.

Per yard super, including excavating and levelling, but exclusive of any concrete foundation:—

	£	s.	d.
Blue Staffordshire paving brick, laid flat, cement grouted	0	7	0
Ditto, chequered brick ...	0	8	0
Ordinary building brick on edge in cement	0	10	0
Pressed brick on edge in cement	0	11	0

	£	s.	d.
Adamantine clinker on edge in cement...	0	12	0
Dutch clinker on edge in cement	0	12	0
Malm bricks on edge in cement	0	6	6
Suffolk white bricks on edge in cement...	0	10	6

If laid in mortar instead of cement deduct 1s. per yard super.

McDougall's Combination Paving Setts.

Heavy pattern, 10in. by 4½in. by 4½in., £12 per 1000, or 6s. per yard super at works.

Footpath pattern, 10in. by 4½in. by 3in., £10 per 1000, or 5s. per yard super at works.

Natural Footpath Flagging.

Yorkshire stone. A1 London. Best quality grey stone ready for use or laid.

2in. per yard super....5 3 or 6 3
2½in. " ...6 6 or 7 6
3in. " ...6 9 or 7 9

Ditto, not so well finished:—

2in. per yard super....4 3 or 5 3
2½in. " ...5 6 or 6 6
3in. " ...5 9 or 6 9

Brown stone, 3d. per yard less than these prices.

Purbeck Stone.

	£	s.	d.
Laid, tooled paving, per foot super...	0	1	6
Laid, rubbed paving, per foot super...	0	2	0
Taking up, redressing, and relaying, per foot super.	0	0	4

Blue Lias Stone.

	£	s.	d.
Laid, 1½in., per foot super.	0	1	2
2in., " ...	0	1	3
3in., " ...	0	1	6
4in., " ...	0	1	8

GRANITE.

Per yard super.

Guernsey Pitching.

	£	s.	d.
5in. × 6in., laid in sand and cement grouted	0	9	0
4in. × 6in.	0	9	0
3in. × 5in.	0	9	6

Aberdeen Pitching.

	£	s.	d.
5in. × 6in.	0	9	0
4in. × 6in.	0	9	6
3in. × 5in.	0	10	0

Aberdeen Granite in Courses.

	£	s.	d.
3in. × 5in., laid in sand and cement grouted	0	17	0
3in. × 7in.	0	19	0
3in. × 8in.	1	1	0
3in. × 9in.	1	2	0
4in. × 5in.	0	15	0
4in. × 7in.	0	17	0
4in. × 8in.	1	0	0
4in. × 9in.	1	0	9
5in. × 5in.	0	10	0
5in. × 7in.	0	12	6
5in. × 8in.	0	15	0
5in. × 9in.	0	15	9

These same prices are applicable to Cornish granite.

1 ton of 6in. granite will cover 4 yards super.

1 ton of 8in. granite will cover 3 yards super.

1 ton of 9in. granite will cover 2½ yards super.

13½ cubic feet of granite = 1 ton.

Grouting granite paving in cement averages 8d. per yard super.

Granite Curb.

Including laying and pointing in cement.

	£	s.	d.
6 × 12, Cornish, per foot run...	0	2	4
6 × 10	0	2	2
8 × 12	0	2	10
6 × 12, Aberdeen	0	2	8
6 × 10	0	2	6
8 × 12	0	3	2

Cornish Channel.

	£	s.	d.
7in. × 5in., dressed on face joints and edges, jointed in cement, per foot run...	0	1	6
Ditto, 12 × 6	0	2	6
Concrete foundation for curb and channel, 6in. thick, per yard super. ...	0	2	6

Cornish Granite.

Placed on rail or ship.

	£	s.	d.
Flagging, not less than 4 square feet area, per foot square	0	1	4
When stones are prepared to suit special widths, per foot square	0	1	7
Curbing, 6in. and 7in., not less than 2ft. 6in. long and 11in. deep, per foot lineal...	0	1	0
Channelling, 12in. × 6in., not less than 18in. long, per foot lineal...	0	1	0
Pitching, 6in. courses, random depths and lengths, per square yard	0	10	0
Cloven granite, in various lengths up to 5ft. and 7in. thick, per foot square ...	0	0	10

Prices of Road Making in 17 Districts.—Initial Cost, including Excavation, Foundation, &c.

Town.	Macadam.		Hard wood.	Soft wood.	Asphalte.	Tarred macadam.	Granite setts.
	Per ton.	Per yard super.					
Aberdeen	—	2/6—3/-	—	—	—	—	15/-
Blackpool	—	5/6	14/-	11/-	—	6/6	—
Bristol	—	3/-	14/-	12/-	—	5/-	—
Cheltenham	—	1/4—1/6	17/-	8/6	—	—	—
Chester	—	2/6	12/9	—	—	—	—
Derby	8/6	—	15/-	10/-	—	3/-	—
East Ham	—	6/-	15/-	—	—	4/0—4/6	—
Hanley	—	3/-	—	—	—	—	15/-
Harwich	11/-	—	11/-	6/6	—	—	—
Leicester	—	4/6	—	—	14/-	—	—
Margate	6/6	3/6	15/-	8/6	—	3/-	15/-
Manchester	—	4/6—5/-	18/6	15/6	—	—	—
Newbury	13/4	—	15/3	—	—	—	—
Penzance	5/6	2/6—3/-	15/6	—	—	3/6—4/-	12/6
Portsmouth	12/-	—	—	—	14/-	—	—
Swansea	—	2/2—2/8	—	—	—	—	—
Worthing...	12/6	—	12/6	10/-	—	—	—

DRAINAGE.

Stoneware Drain Pipes.

At works. Albion Clay Company.

Diam.	Per yard. £	s.	d.
2in.	0	1	1½
3in.	0	1	2
4in.	0	1	6
5in.	0	1	10½
6in.	0	2	3
7in.	0	2	9¾
8in.	0	3	4½
9in.	0	3	9
10in.	0	4	6
12in.	0	6	4½
15in.	0	10	6
18in.	0	15	9
21in.	1	4	0
24in.	1	11	6

Drain Interceptors.

Diam.	Each.		
4in.	0	8	0
6in.	0	11	0
9in.	0	17	0

Manhole Interceptors.

4in.	0	11	0
6in.	0	16	0
9in.	1	4	0

Glazed Stoneware Drains.

Best quality pipes, laid and jointed with gasket and cement, including all trench work and refilling and ramming.

Size of drain. Inches.	Average depths and cost per foot. 2ft. 6in.	3ft.	6ft.	9ft.	12ft.
4	1/-	1/5	1/10	2/6	3/-
6	1/3	1/8	2/-	2/9	3/3
9	1/8	2/3	2/9	3/3	3/9
12	3/-	3/6	4/2	5/-	5/6

Ditto, surrounded with 6in. Portland cement concrete, 6 to 1.

Size of drain. Inches.	Average depths and cost per foot. 2ft. 6in.	3ft.	6ft.	9ft.	12ft.
4	1/8	2/1	2/6	3/2	4/-
6	2/1	2/6	2/10	3/7	4/9
9	3/-	3/7	4/1	4/7	6/-
12	4/10	5/7	6/6	7/4	8/6

Agricultural Drain Pipes.

	£	s.	d.
2in., per 1000	2	0	0
3in.	4	0	0
4in.	5	10	0
6in.	10	10	0

Subsoil Roadside Drains.

Trench 2ft. deep, and box drain of old bricks, drain tiles, or flat stones, and trench filled in with coarse stones.

	Per foot lineal.		
Drain 4in. square internally	0	1	6
6in. × 4in.	0	1	9
12in. × 6in.	0	2	5

Brick Barrel Drains.

Per foot run.

4½in. brickwork.

	In mortar. s.	d.	In cement. s.	d.
9in. drain	1	8	2	2
12in.	2	0	2	6
15in.	2	6	3	0

9in. brickwork.

	s.	d.	s.	d.
15in.	5	6	6	0
18in.	5	9	7	0

Rendering in cement, 4d. per foot super.

For excavation and filling add 2d. per foot run for each foot deep.

Cast Iron Drains.

Laying pipes in 9ft. lengths, making lead joints, and providing all lead and tackle, but not pipes.

	£	s.	d.
4in., per foot lineal	0	0	4
5in.	0	0	6
6in.	0	0	8
9in.	0	0	10
10in.	0	1	0
12in.	0	1	3

For excavation and refilling add 2d. per foot lineal for each foot deep.

Cast Iron Socket Pipes.

Approximate price at Messrs. Jordan and Co.'s works in Newport for tested and coated socket pipes in 9ft. lengths.

Inches.	Weight. Cwt.	Qr.	Lb.	Per ton. £	s.	d.
3	1	0	12	6	0	0
4	1	2	7	5	17	6
5	2	0	7			
6	2	2	14			
7	3	0	0	5	15	0
8	4	0	0	5	13	0
9	4	2	0			
10	5	0	0			
12	7	0	0			

RETAINING AND BREAST WALLING.

Brick.

Of Staffordshire bricks in cement mortar, 2 to 1. Per rod reduced, including profit :—

	£	s.	d.
Labour, mortar, and scaffolding	9	16	8
Work complete, less pointing and coping	31	15	0

In selected stock bricks, and stone lime mortar :—

	£	s.	d.
Labour, mortar, and scaffolding	6	12	0
Work complete, less pointing and coping	16	0	0
If in blue lias mortar add...	0	10	9
If in cement mortar add ...	2	15	0

In grizzles :—

	£	s.	d.
Labour, mortar, and scaffolding	6	5	0
Work complete, less pointing and coping	14	10	0

Add in each case :—

	£	s.	d.
For circular work, per yard super...	0	2	3
For battered face, per yard super...	0	0	9
For rendering in cement, per yard super...	0	1	6
Blue Staffordshire coping in cement, per course, per yard run	0	1	0
Stock brick on edge in cement, per course, per yard run	0	0	8
Pointing flat joints in mortar, per yard super...	0	2	2½
Pointing flat joints in cement, per yard super...	0	2	11½
Tuck pointing and staining brickwork, per yard super	0	2	3

Stone.

Price of stone, 6s. per yard cube, which price must vary very considerably.

	£	s.	d.
Random work in walls, 20in. thick, stone roughly squared, laid in lime mortar, per yard super...	0	14	0
Ditto, 18in. thick	0	12	0
Ditto, 12in. thick	0	10	0
Ditto, with scappled face and drafted margin :—			
20in. thick, per yard super.	0	16	6
18in. " " ...	0	14	6
12in. " " ...	0	13	0

	Per yard super.		
	£	s.	d.
Pointing flat in mortar ...	0	1	0
" " cement ...	0	1	6
" tuck in mortar ...	0	1	5
" " cement ...	0	2	0

IRONWORK, &c.

	£	s.	d.
Cast iron gully gratings, per cwt....	0	10	0
" bends, per cwt....	0	11	6
" junctions, per cwt.	0	11	6
" ferrules, per cwt.	0	10	0
" manhole covers, per cwt.	0	10	0
" stop tap boxes, small, each ...	0	1	3
" stop tap boxes, medium, each...	0	3	6
Wrought iron gully gratings, 24in. × 16in., each ...	1	5	0
Navvy shovels, riveted eye handles, per doz.	0	19	6
Navvy spades, per cwt. ...	2	0	0
Navvy picks, solid eyes, per cwt.	1	3	0
Sledge hammers, steel faced, each	0	8	6
Ash handles for ditto, each	0	1	0
Hand hammers, steel, each	0	3	0
Ash handles for ditto, each	0	0	7
Stone breakers' hammers, per doz.	0	10	0
Ash handles, per doz. ...	0	5	0
Chisel steel, per lb.	0	1	0
Screw bolts and nuts, per lb.	0	0	3
Wrought iron pile shoes, per lb.	0	0	3
Rolled steel joists, per cwt., from	9s. to 11s.		
Wrought and riveted steel plate girders, per cwt., from	14s. to 18s.		

	£	s.	d.
Sluice valve boxes, 3½in. diam. opening, each ...	0	6	6
Hydrant boxes, top 13in. × 9in., depth 7½in., each ...	0	9	0

Tree Guards, painted one coat.

	£	s.	d.
6ft. high, wrought iron, 1¼ × ¼, vertical bars 2in. apart, pronged feet:—			
Diameter, 9in., per doz. ...	5	0	0
12in.	6	10	0
15in.	7	10	0
18in.	9	5	0
21in.	10	10	0
24in.	11	10	0
6ft. high, 20in. diam. at top, 16in. at bottom, and 12in. at centre, per doz.... ...	12	0	0

Fencing.

Continuous wrought iron flat-bar fencing, 4ft. high, standards 4ft. in ground, 3ft. apart, with ⅝in. top round bar rail, suitable for fencing in fields.

Per yard lineal	0	2	8

A plain fence for drives, 3ft. 9in. high, standards 4ft. apart, 5⅝in. diameter wrought iron round horizontals, cast iron standards 1ft. 6in. in ground.

Per yard lineal	0	7	6

Pavement Lights.

	Per foot square.		
Convex lens lights	0	5	0
Corrugated lens lights ...	0	5	0
Studded lens lights	0	6	0
Semi-prism lights ...	6s. to 7s. 6d.		
Heavy lights for carriage traffic, from	9s. to 13s.		

STONE BREAKERS.

Baxter's Patent Improved Compound Toggle Knapping-motion Stone Breaker. **On wheels with horse shafts and steel screen.**

Price.

Size at the mouth of machine.	Price, delivered at Leeds station.	Quantity broken to [not larger than 2¼in.] per hour.		Approximate weight.
		Approximate.	Guaranteed.	
	£	Tons.	Tons.	Tons.
6 × 4	76	2¼	1½	1¾
8 × 5	95	4	2½	3
12 × 6	125	6	4	4
12 × 8	158	7½	5	6
14 × 8	180	9	6	7¾
16 × 9	208	12	8	9
20 × 9	260	15	10	10
20 × 12	290	16	11	11¼
24 × 12	355	22	15	17½
24 × 15	395	23	16	19

Baxter's Patent Automatic Screening and Loading Machine, with Knapping-motion Stone Breaker. **For fixing on concrete foundations.**

Rock Broken per Hour.

Size at the mouth of machine.	Quantity broken to [not larger than 2¼in.] per hour.		Price on rail, Leeds.	Approximate weight.
	Approximate.	Guaranteed.		
	Tons.	Tons.	£	Tons.
12 × 6	6	4	170	3¾
12 × 8	7½	5½	200	5¾
14 × 8	9	6	225	7
16 × 9	12	8	260	8¼
20 × 9	15	10	315	9¾
20 × 12	16	11	350	11¼
24 × 12	22	15	410	18
24 × 15	23	16	455	19¼

Blake-Marsden Stone Breaker.

Size of machine at mouth.		Approx. product per hour, broken stone. Tons.
10 × 8	... ...	4
12 × 8	... ...	5¼
12 × 9	... ...	5½
15 × 8	... ...	6¼
15 × 10	... ...	7½

Power required. N.H.P.	Price on feet.	Wheels, horse shafts, automatic screening apparatus, and belts complete, extra.
	£	£ s. d.
3	105	30 18 0
3	113	30 18 0
3	120	31 8 0
5	135	38 13 0
6	150	39 13 0

Mason's Improved Lever Stone Breaker.

Size at mouth.	Power required.	Quantity broken per hour. Tons.	Price with screen. £
8× 4	2-man	2	47
10× 7	2-horse	4	70
12× 8	3 "	5	90
15×10	4½ "	7½	133
16×10	5½ "	8½	156
20×10	7 "	10½	186
24×10	8½ "	11	216
24×12	9 "	12	226

STREET GULLIES.

Brick in cement gullies and catchpits rendered in cement. Labour, materials, and connection.

	£ s. d.
Without grating or collaring	2 18 6
Including granite collaring	3 13 6
Including C.I. grating ...	4 3 6

Circular or rounded angle stoneware street gullies and catchpits.

Albion Clay Company, Ltd.

Diam.	Depth. ft in.	Price. £ s. d.
6in.	2 0	0 6 0
9in.	2 0	0 10 0
"	2 6	0 12 6
12in.	2 0	0 15 0
"	2 6	0 17 6
"	3 0	1 0 0
15in.	2 0	1 0 0
"	2 6	1 3 6
"	3 0	1 7 0
"	3 6	1 11 0
18in.	2 0	1 5 0
"	2 6	1 10 0
"	3 0	1 16 0
"	3 6	2 2 0
"	4 0	2 10 0
"	4 6	3 0 0

Clark's Patent.

Inlet dimensions.	Treble trapped. £ s. d	Double trapped. £ s. d.
30 × 18	5 0 0	4 15 0
24 × 16	4 0 0	3 15 0
20 × 15	2 0 0	1 17 0
18 × 12	1 13 0	1 10 0
Cast iron gully gratings, per cwt.		0 9 0
Wrought iron gully gratings, 24 × 16, each...		1 5 0

IMPLEMENTS.

Watering Cart.

With springs, box distributor, wrought iron tank. Best finish, painted.

	£ s. d.
200 gallons	28 0 0
250 "	30 0 0
300 "	32 0 0

ROTARY WATERING MACHINE ON TWO WHEELS.

	£	s.	d.
Wood tank	35	0	0
Steel tank	37	0	0
Balancing gear to carts, extra	3	0	0

WATERING VANS.

Four wheels, steel body, box distributor, best finish, painted, with brake.

	£	s.	d.
350 gallons	45	0	0
400 „	48	0	0
450 „	53	0	0

Vans with rotary watering machine.

	£	s.	d.
350 gallons	52	0	0
400 „	55	0	0
460 „	57	0	0
Hand watering cart, to hold 75 gallons...	14	0	0

BROOM SWEEPERS.

Best makes from £30 to £35 each.

HORSE ROAD SCRAPERS.

6ft. wide from 12 to 16 guineas.

HAND ROAD SCRAPERS.

All iron and steel.

Scrapers.	ft.	in.	£	s.	d.
8, full width...	2	8 ...	4	4	0
10	3	4 ...	4	15	0
12	4	0 ...	5	10	0

IRON MUD CARTS.

	£	s.	d.
150 gallons capacity	25	0	0
200 „	29	0	0
260 „	32	0	0
280 „	33	0	0

WOOD MUD CARTS.

Oak framing, tipping gear, fitted with iron cover.

	£	s.	d.
350 gallons capacity	25	0	0
400 „	28	0	0

TIP WAGONS.

Suitable for dust or mud. With iron cover.

	£	s.	d.
To hold 1½ cubic yards ...	50	0	0
„ 2 „ ...	55	0	0
„ 2½ „ ...	60	0	0

INDEX.

NEW ILLUSTRATED CATALOGUES POST FREE ON APPLICATION.

ALBION CLAY CO., LTD.

MANUFACTURERS.

ALBION WORKS.	CHIEF LONDON OFFICE.
WOODVILLE, BURTON-ON-TRENT.	38, VICTORIA STREET, WESTMINSTER, S.W.
Works Telegrams—"ALBION WOODVILLE."	*London Telegrams*—"SEWERAGE, LONDON."
,, *Telephone*—14, SWADLINCOTE.	,, *Telephone*—5536, GERRARD.

TRADE

MARK.

TRADE

MARK.

Quotations given for a l Sizes of Pure Stoneware Pipes by Van, cr f.o.r. to any Station, and f.o.b. to any Port in England.

SYKES' PATENT STONEWARE STREET GULLY,

ALL SIZES.

DEEP SEAL.
DEEP CATCHPIT.
EASY ACCESS.
OUTLET WELL BELOW CROWN OF ROAD.

OUTLET A SUPPLIED TO ORDER

Stoneware,
Price 60s., with Iron Grate, delivered in lots of 10
to any Railway Station in England.
Awarded Medal of
The Sanitary Institute,
1899.
Also made with access to pipes from inside of Gully.
Made also in Iron

SYKES' NEW PATENT PIPE JOINT

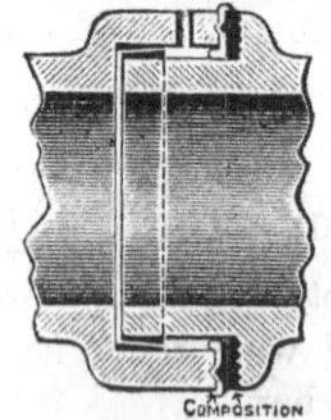

YKES' PIPE JOINTED.

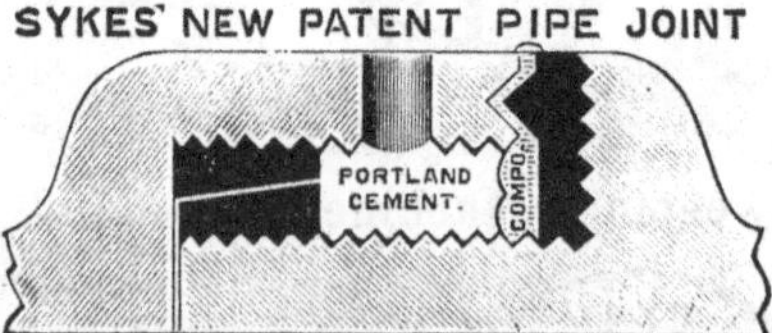

ENLARGED SECTION (JOINTED).

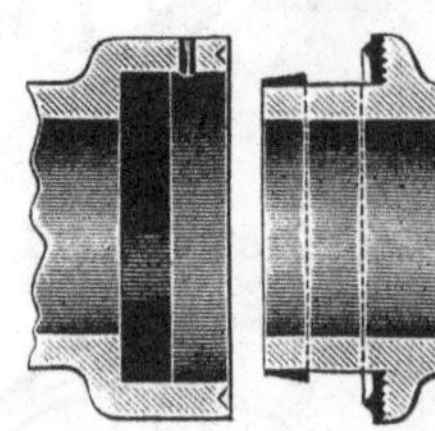

BEFORE JOINTING.

PATENT PARAGON PIPE.

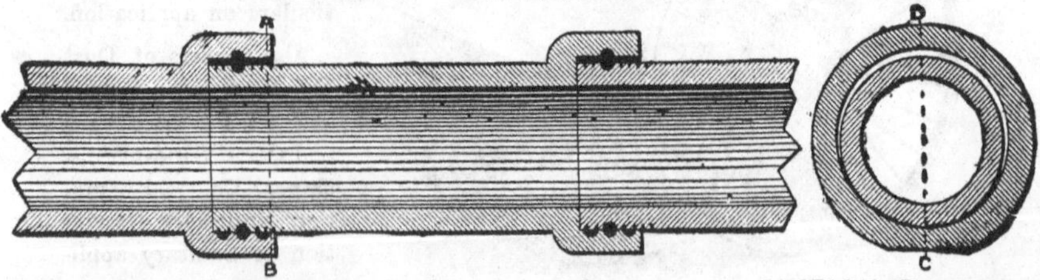

LONGITUDINAL SECTION THROUGH C.D. SECTION THROUGH A.B.

True Alignment of Invert. Spigot cannot Drop. Easily and Quickly Laid. Moderate Cost.

DETAILS RELATING TO ROAD CONSTRUCTION, TABULATED BY THE AUTHOR, FROM REPLIES RECEIVED FROM SOME LEADING ENGINEERS, EXPERIENCED IN THIS IMPORTANT BRANCH OF CIVIL ENGINEERING, AND TO WHOM THE AUTHOR IS INDEBTED FOR THE VALUABLE INFORMATION.

Authority.	With what description of Materials are your Roads Maintained?	What is the Approximate Average Thickness Worn per Annum, and Description of Traffic over same?	What is the Approximate Initial and Annual Cost of the Materials you use?	Do you adopt a Road Scarifier or Hand Labour for Breaking your Roads? State the advantages of the former, if any.	Is your Macadam broken by Hand or Machine? Do you find the Life of the Material reduced by one or the other?	What save, or otherwise, do you consider there is in Scavenging Wood and Asphalt Roads, as compared with Macadam Roads?	Do you use Salt Water for Street Watering? If so, with what advantage, or otherwise?
J. W. Cockrill, Great Yarmouth	Macadam—Guernsey granite, Rhenish basalt, and Cherbourg quartzite. Wood—Jarrah and Memel.	Jarrah has worn 1/8in.	—	Aveling and Porter's road scarifier in use. Road is much better broken up with a scarifier than by hand labour, at one-tenth cost.	Prefer hand-broken, but have difficulty in getting it.	Wood and asphalt require to be kept cleaner than macadam. In my case I consider it costs double to scavenger a paved street to what the macadam does.	We use salt water for street watering. It reduces the cost of maintenance of light macadam roads very considerably; it saves slightly, in my opinion, on heavier metalled roads; and costs a little more, perhaps, to keep wood paving clean.
Thos. Henry Yabbicom, Bristol	Macadam—Carboniferous limestone (local), Cornish Elvan, Wick, Tytherington, Clee Hill. Wood—Karri, Jarrah, prismatic oak, beech, red Swedish deal.	Macadam wears according to description and quantity of traffic. Wood ditto ditto. Some wood-paved carriage-ways get as much, and, in some instances, more than the sett-paved carriageways.	Macadam. Initial cost per yard super, 3s. Annual cost of maintenance varies as the quantity and description of traffic. Hard wood. Initial cost per yard super, 14s. (9720 yards super laid in 1895 and since). No repairs on wood paving executed in hard wood up to present time. Soft wood. Initial cost per yard super, 12s. (151,247 yards super laid in 1878 and since). Annual cost of maintenance only per yard super, 6d. The high cost of soft wood is due to the exceptionally high cost of Swedish timber at the present time; as a consequence very little of this class of timber is being laid in this city at present.	Yes. Aveling and Porter, only used on large areas. The cost varies according to the description of stone used for the macadam.	Both. Hand-broken has a longer life than machine.	The amount of scavenging depends very largely upon the amount of traffic. Where the traffic is the same, a macadam road costs considerably more to cleanse than a wood-paved road.	No.
Wm. Tyack, Aberdeen	Macadam—Granite. Tar macadam—Granite.	Principal streets paved with granite setts. Macadam used only where traffic is light. Average wear about 1in.	Ordinary macadam, including hand-set bottom. Initial cost, 2s. 6d. to 3s. per square yard. Annual cost about 4d. Tar macadam. Initial cost 4s. to 4s. 6d. per square yard.	The scarifier with three men will do as much in one day as ten men will do in a week by hand.	Principally by hand. Machine-broken metal does not last so long as hand-broken.	—	No.
Arthur C. Turley, Canterbury	Macadam—Guernsey granite, gravelstone flints. Wood—Jarrah blocks.		—	The saving effected by using a scarifier as against hand labour is enormous.	Hand-broken stone is certainly more durable than machine-broken, as it binds together better.	Cost of scavenging ordinary macadam is probably three times greater than that of wood or asphalt.	No.
Joseph Lobley, Hanley	Macadam—All kinds of granite, &c., principally Rowley Regis basalt.	Macadam, 3in.	Macadam, 4s. per square. Granite setts on concrete foundation, 15s. per square yard.	Half the cost saved.	Both. Hand-broken is the better of the two.	The cost of scavenging paved streets is less than half that of macadamised.	No.
Frank Roberts, Worthing	Macadam—Granite and flints. Wood—Duffy's Patent. Asphalt—Val de Travers.	Macadam Hard wood } Varies. Soft wood	Macadam, 12s. 6d. per ton. Hard wood, 12s. 6d. per yard on own concrete. Asphalt, 10s. per yard on own concrete.	Yes.	By hand.	Very great.	No.
T. de Courcy Meade, Manchester	Macadam—Penmaenmawr, from Messrs. Brundrit and Co.'s and Messrs. Darbishire and Co.'s quarries. Wood—Creosoted English beech and Australian Karri and Jarrah.	Unable to give this. Traffic in centre of town is very heavy.	Macadam. Initial cost, 4s. 6d. to 5s. per square yard. Annual cost of repair and maintenance varies according to position and traffic. Hard wood. Initial cost of Jarrah and Karri is about 18s. 6d. per square yard. Unable to give annual cost of repair and maintenance as yet. Soft wood. Initial cost of creosoted English beech is about 15s. 6d. per square yard. Annual cost of repair and maintenance is about 1s. 1½d. per square yard.	We employ hand labour.	All our macadam is purchased hand-broken.	Cost not separated.	No.
Chas. Brownridge, Birkenhead	Macadam—Penmaenmawr. Wood. Asphalt—Tar macadam.	—	Depends entirely upon position and local circumstances.	Hand labour.	Both. I find that with modern machinery there is no practical difference in the stone broken by hand and machine.	Very considerable, if they are properly done.	Yes.
John Ward, Derby	Macadam—Leicestershire granite, broken to 2½in. gauge, and now trying best slag, same gauge, on bye streets. Wood—Creosoted red deal blocks grouted with pitch, small length of beech, and small length of Jarrah. Asphalt—Tarred macadam, composed of broken slag, tar, and creosote oil.	—	Macadam, 8s. 6d. per ton, about 2s. per square yard for 3in. coat laid. Hard wood, about 15s. per square yard (including concrete). Soft wood, £9 per 1000, 9in. by 6in. by 3in., about 10s. per square yard, ditto. Asphalt. Tarred macadam, 3s. per square yard laid. Prices would now be slightly higher than above. No annual cost kept of repairs. Wherever this is done, the figures are often most misleading. Depends entirely on class of traffic and disturbances of surface.	Both. Scarifier, Morrison's; makers, Aveling and Porter. Great saving compared with hand power.	Machine.	Considerable.	No.
Wm. J. Newton, Accrington	Macadam—Granite, Welsh. Wood—Small portion Australian hard wood.	Hard wood. Only recently laid, five years ago; traffic heavy in centre of town.	Hard wood. No expenditure since we had the setts some five years ago.	We have always used spikes for the rollers. Aveling and Porter's make.	We always use 2in. hand-broken stone.	—	No.
Wm. Bolton, Wigan	Macadam—Welsh granite. Wood—Red gum and American white oak, 9in. by 5in. by 3in.	Wood blocks in one street, only just been laid.	—	No.	Machine-broken. I do not think there is any material difference.	—	No.
Geo. Walkley, Llanelly U.D.C.	Macadam—Limestone and granite.	Macadam. Vary, according to traffic, from 1in. to 3in.	—	Have used a scarifier, which is a saving of at least 50 per cent., taking everything into consideration. It is Aveling and Porter's make—a "Morrison."	We buy all our stones broken. I should say these are principally broken by machine. There should be no difference so long as the gauge is right.	—	If there was an advantage in using salt water, it would be too costly.
H. Gilbert Wyatt, Grimsby	Entirely macadam—chiefly Belgian granite. Am now putting down experimental lengths of Rhenish basalt. All stone 1½in. gauge.	Macadam. Cannot state. A road done 4in. thick is trimmed once in its life, and when it wears down to about 1½in. thick is made up to 4in. again.	Cannot state definitely. Last year spent £2950 to "materials," which includes flags, kerbs, channels, &c. Length of roads about 50 miles.	At present, hand labour. Have recently accepted a tender for an additional roller with scarifier, and the application is now before the Local Government Board. Cannot state the saving from experience. A. E. Collins, Esq., City Engineer of Norwich, some years ago wrote a paper on the cost and economy of scarifying for the Municipal and County Engineers.	Machine-broken.	Great saving.	No.
J. Matthew Jones, Chester	Macadam—Penmaenmawr basalt. Wood—Jarrah and deal.	In heavy traffic macadam streets repaired two or three times a year. Hard wood not been long enough laid. Soft wood requires repair in five years; perhaps total renewal in ten years.	Macadam, 3s. 6d. per yard, 6in. thick. Cost of maintenance not separated between light and heavy traffic. Hard wood, 12s. 9d. per yard. No repair item up to the present. Soft wood, 8s. 9d. per yard. No foundation price included.	No.	By hand and machine. I think it is better to use hand-broken.	Owing to droppings, &c., lying on wood pavements, we have to employ special pickers-up. In case of macadam, general cleaning is less expensive, but when rain falls four times the quantity of mud has to be removed from this surface.	No.
A. Jh. Evans, Luton	Macadam—Leicester granite, field-picked flints, furnace slag (broken).	Macadam wears about ¾in. to 1in. on most roads. Heavy town traffic about 1½in.	Macadam, granite, 10s. 6d. to 12s. per ton at station; £800 per annum. Slag, 6s. per ton at station; £200 per annum. Flints, 6s. to 7s. per cube yard on roadside; £400 per annum.	Use Morrison's scarifier. About ½d. per super yard.	Machine.	—	No.
A. W. Campbell, East Ham	Macadam—Guernsey granite. Wood—Jarrahdale Jarrah. Asphalt—Tarred macadam.	—	Macadam, initial 6s. per yard. Hard wood, 15s. per yard. Tarred macadam, 4s. 6d. per super yard.	Yes; about half as costly as stocking by manual labour.	By hand, and would have none other.	—	No.
John Gamage, Dudley	Macadam—Rowley ragstone, quarries in the borough.	Macadam wears 3in. on main roads; both heavy and light traffic.	Macadam, 7s. 6d. per ton delivered on to roads; £2200.	No.	Some by hand, and some by machine; the life of the hand-broken is longest.	Have no paved streets except on tramway route.	No.
Henry Ditchan, Harwich	Macadam—"Penlee" and "Quenast" stones; also broken Kentish flints. Wood—Jarrahdale and fir.	We have no exceptionally heavy traffic over our roads, railway vans and coal trolleys, carrying about two tons each, being the heaviest. As to annual wearing away:— (1) Macadam—If kept well watered, so as to prevent disintegration, the wear is comparatively slight, certainly not more than ½in. per year. (2) Hard wood—I have some Jarrah blocks which have been down eight years, and they are not worn down ½in. (3) Fir—Goes very quickly, and requires to be renewed in seven to eight years. Size of blocks used, 3in. by 9in. by 4½in. deep.	Macadam, 1¼in. gauge Penlee stone, 11s. per ton. Coarse Penlee chippings, 9s. per ton. Quenast stone, 10s. 6d. per ton. Hard wood, 11s. per yard, all materials and labour. Soft wood, 6s. 6d. per yard, all materials and labour.	Dutty's preferred. Cost per yard, ½d.	Hand-broken.	There is far less mud on hard wood paving to pick up after the horse brooms have been over it, and there is also less dust in summer.	Yes. We have a central station and reservoir, to which the water flows by gravitation. It is then pumped up to a service tank (2000 gallons), which fills the water vans. Results, excellent.

The material originally positioned here is too large for reproduction in this reissue. A PDF can be downloaded from the web address given on page iv of this book, by clicking on 'Resources Available'.

DETAILS RELATING TO ROAD CONSTRUCTION, TABULATED BY THE AUTHOR, FROM REPLIES RECEIVED FROM SOME LEADING ENGINEERS, EXPERIENCED IN THIS IMPORTANT BRANCH OF CIVIL ENGINEERING, AND TO WHOM THE AUTHOR IS INDEBTED FOR THE VALUABLE INFORMATION (*continued*).

Authority.	With what description of Materials are your Roads Maintained?	What is the Approximate Average Thickness Worn per Annum, and Description of Traffic over same?	What is the Approximate Initial and Annual Cost of the Materials you use?	Do you adopt a Road Scarifier or Hand Labour for Breaking your Roads? State the advantages of the former, if any.	Is your Macadam broken by Hand or Machine? Do you find the Life of the Material reduced by one or the other?	What save, or otherwise, do you consider there is in Scavenging Wood and Asphalt Roads, as compared with Macadam Roads?	Do you use Salt Water for Street Watering? If so, with what advantage, or otherwise?
J. Herbert Morris, Godalming ...	Macadam — Guernsey granite, Mendip limestone, Farnham flints, Bargate sandstone (local). We have about a mile of Aberdeen and Guernsey granite pitching.	Macadam. No accurate observations of the thickness worn per annum have been made. Our traffic is considerable, there being several large mills, to and from which heavy loads are hauled. There is also the usual trade traffic of a country town fairly busy for its size.	Macadam, granite, 18s. 6d. ton laid, life three years. Mendip limestone, 11s. 6d. ton laid, life two years. Bargate and flint, 10s. ton laid, life 1½ years. On roads of equal traffic. No reliably accurate observations have been made.	Voysey and Hosack's scarifier on most roads, 1d. per yard super cost. Average cost of hand picking 1½d. per yard, and work not so effective. Roads can be brought to better slope, and at least 5 per cent. new metal saved by using scarifier.	The granite and limestone is all broken by machine away. Local stone is broken by hand, as machine-breaking is distinctly deteriorating.	Have no wood or asphalt. With the same traffic, the cost of scavenging granite pitching is not half that of macadam.	No.
J. Siddalls, Tiverton	Macadam—Chiefly local limestone, a little Leicestershire granite. Asphalt—I laid down one street in tar macadam about six years ago for second-class traffic. It has answered admirably, and has cost practically nothing in repair up to this time.	Macadam varies with the traffic. With fairly heavy traffic the limestone will wear 3in. a year; the granite not more than 1in.	—	Have only just begun to use a scarifier, but there is an enormous saving in labour and materials.	By hand almost entirely. I consider the softer stones, such as limestone, much depreciated by machine-breaking.	I cannot give figures, but there is no question of the great saving of scavenging an impervious surface which rapidly dries over that of an ordinary macadam road which yields mud at every shower.	No.
G. H. Pickles, Burnley	Macadam—Threlkeld granite (our roads are principally paved with stone or granite). Wood—Jarrah and beech (6in. concrete, 1in. sand, 5in. blocks). Asphalt — Tar asphalt of broken granite coated on tar when hot, surface finished with 1in. of granite chippings rolled in.	—	—	We do not use a road scarifier, the usual points are inserted in wheels of road roller.	By machine.	Our wood-paved road is not yet scavenged with the remainder of the roads, but has separate attention.	No.
W. J. Ebbetts, Acton	Macadam—Guernsey granite. Wood—Creosoted fir blocks.	Ordinary traffic over macadam. Wear of soft wood not appreciable—ordinary traffic.	Macadam, 22s. 6d. per yard cube, annual cost. Soft wood, 12s. per yard super, first cost.	No.	By machine—no comparison has been made.	This has never been gone into.	No.
A. E. Nichols, Folkestone	Macadam — Guernsey granite, Kentish rag, flint. (Material in accordance with situation and work.) Wood—Partly soft wood and partly Jarrah. Asphalt.	Macadam worn on main roads 1½in. to 2in. in two to three years. Hard wood worn practically nothing. Soft wood rots rather than wears.	Macadam, granite, broken, 12s. 10d. per ton on quay. Rag, 7s. 6d. per load, delivered. Flint (broken), 9s. 6d. per cube yard, delivered. Hard wood, Jarrah, £10 15s. 6d. per 1000, on rail to Folkestone. No repairs due to wear. Soft wood, rotten after three years, and am pulling out and replacing with hard wood or York setts.	Labour.	Machine for granite, hand for rag and flints. (Hand-broken undoubtedly lasts longer, and does not grind up so much when rolling).	—	No.
Thos. Cookson, Preston	Macadam—Granite.	Macadam worn about 4in.	Macadam. About £1000.	Don't use scarifier.	Both; very little difference.	No experience in wood and asphalt roads.	No.
R. Read, Gloucester...	Macadam—Basalt stone from Clee Hill Dhu Stone Co., Ludlow, Shropshire. Wood—1250 square yards of Baltic red deal at the Cross, in centre of city, cut off 9in. by 3in. deals 5in. deep.	Macadam roads in a town do not wear evenly like they do in the county districts, owing to the numerous openings for various purposes to which they are constantly subjected, which makes them wear in holes which vary in depth according to the nature of the foundation and the care which is taken in filling up excavations. Soft wood. This has only been down five or six years, and has worn from ⅛in. to ¼in. in thickness. This wood paving is exposed to a continuous and mixed traffic of all kinds concentrating from four directions in the very heart of the city.	A macadam road, with hand-pitched oolite limestone foundation, 9in. thick, and 4in. of Clee Hill stone, would cost 3s. a square yard to construct. We have 40 miles of road, and the average cost of maintenance is £100 per mile per annum. Soft wood, 10s. 6d. per square yard. No expenses for repairs. I think all roadmakers are agreed that, although macadam is the cheapest in first cost, it is by far the dearest in the long run.	No scarifier. From what I have seen of them in work, they make very little impression on a hard road made with Clee Hill stone; and a soft limestone road they drag up by the roots.	By hand. I consider hand-broken stone increases the life of a road by 20 per cent., but a smoother road can be maintained with machine-broken stone than with hand-broken.	A road made with carboniferous limestone will make three times the mud that a road made with Clee Hill stone will make, a wood-paved road will make about one-third the mud a Clee Hill stone road will, and an asphalt road one-quarter.	No.
Albert Latham, Margate...	Macadam—Flints and quartzite. Wood—Creosoted fir, run in with asphalt, and cement grout on 6in. of concrete.	Principal macadam roads wear 3in.; others, 1in. to 2in.; wood, ¼in.	Macadam, first cost, flints, 2s. 6d. yard super; first cost, quartzite, 4s. 6d. yard super. Soft wood, including foundation, 10s. yard super.	No.	Hand-broken, as a rule.	Situated as our paved road is, between the harbour, gasworks, and waterworks, and being a sunless street, the cost was four times as much to cleanse the road when it was macadamised, compared with the present wood.	Yes. Have used sea water on a few roads for some years with good results. I am now carrying out a pumping scheme so that the salt water may be used throughout the town. One sprinkling of salt water is as effective in laying the dust as three of fresh. A crust is formed on the surface of the road which lays the dust for some time after the road is dry.
G. Bell, Swansea	Macadam—Syenite limestone and native sandstone. Asphalt — Some tar macadam used for crossings.	—	Macadam. Syenite, 2s. 8d. per square yard initial cost, including excavation, foundation, metalling, and rolling. Annual cost per square yard, 3·15d. Limestone, 2s. 3d. per square yard initial cost, including as above. Annual cost, 1·93d. per square yard. Sandstone, 2s. 2d. per square yard initial cost, including as above. Annual cost, 1·78d. per square yard.	Both road scarifier and hand labour used; Morrison's scarifier. Consider that scarifying costs about half as much as hand labour.	Chiefly machine broken. The native sandstone and some limestone is broken by hand. No difference in life has been observed.	No records. Consider there is a great saving.	No.
S. J. Lea Vincent, Newbury	Macadam—Black Welsh granite. Wood—Karri hard wood, which I have found to be better than Jarrah.	Macadam. Difficult to say exactly, but, judging from wear by manhole covers, &c., should say ¾in. to 1in.	Macadam. Granite, 13s. 4d. a ton delivered. Labour and steam rolling about 3s. 9d. a yard cube. Hard wood, 15s. 8d. yard super, on 9in. 5 to 1 concrete.	No.	Hand.	It costs us one-fifth to scavenge wood.	No.
Thos. Satuk, Worcester	Macadam—Clee Hill and Rowley Regis stone. Wood—Oak, some Jarrah.	Macadam, ¾in. Hard wood, ⅛in.	Macadam. Initial cost about 3s. 6d. per super yard, and annual cost with our traffic averages about 8d. per yard. Hard wood. On 6in. concrete 14s. per yard. About 10d. per yard.	By hand, but am just introducing a Morrison's scarifier.	Broken by hand and machine. Hand-broken seems to last somewhat longer.	The cost of scavenging wood pavement about one-half that of macadam.	No.
Wm. Mackison, Dundee...	Macadam—Blue Whinstone.	—	Macadam. Initial cost, exclusive of bottoming, 1s. 6d. per super yard. Maintenance approximately, including renewal, 6d. per square yard. These figures apply to streets carrying fairly heavy traffic.	Labour. A scarifier would be preferable, and a saving would be effected by its adoption. Have recently experimented with scarifier in excavating track for tramways in macadamised road, and found a saving in labour of at least 60 per cent.	Presently by hand. Hand-broken metal has longer life, as the stones have sharper fracture and are more solid and cubical than is the case of machine-broken metal, which is often flaky, rounded on edges, and injured by the crushing.	—	No.
Joseph Hall, Cheltenham	Macadam—Local oolite, mountain limestone from near Bristol, Clee Hill basalt. Wood—700 square yards Jarrah just laid.	Average cannot be given, as we have many very wide roads which make an average misleading.	Macadam. Local stone about 10d. to 1s.; limestone, 1s. 4d. to 1s. 6d.; Clee Hill, 2s. Hard wood, 17s. per square yard.	Hand labour at present. Am considering the use of a scarifier.	Machine. No appreciable difference.		Haven't any.
Donald Cameron, Exeter	Macadam—Flint and limestone mixed; two-thirds flint, one-third limestone.	—	Macadam. Annual cost, including scavenging, £126 per mile. Flints cost 5s. 5d. a ton, limestone 5s. 4d. a ton, breaking 1s. 3d. a ton. Carting averages 1s. 2d. a ton. A ton of metal covers about 5 yards.	No.	Both.	—	No.
E. George Mawbey, Leicester ...	Macadam—Leicestershire granite, and syenite. Wood—Creosoted deal, Karri, and Jarrah. Asphalt—Ordinary tar macadam.	Heavy traffic, both slow and quick omnibuses, motor cars, &c., over macadam, hard wood, and soft wood. Traffic over asphalt similar to above in some instances, but in others quiet side streets, with offices occupied by lawyers, &c.	Macadam. Initial cost, say, 4s. 6d. a yard super. Annual cost depends upon traffic, and varies with each street. Hard wood. Initial cost, say, 15s. to 18s. a yard super. Not separated in our accounts. Soft wood. Initial cost, say, 10s. to 12s. a yard super. Not separated in our accounts. Asphalt. Tar macadam, 5s.	We do use a scarifier, but the greater portion is done by hand labour. Have not had sufficient experience of the machine scarifier to enable a correct estimate to be formed, but there is no doubt a considerable saving.	Machine-broken. Hand-broken is, of course, the best.	Wood and asphalt roads are cheaper to scavenge than macadam roads.	No; too far away from the sea.
F. Chambers, Goole...	Macadam—Heavy traffic, Durham whinstone; moderate traffic, Belgian granite. Wood—I have tried three years to secure a trial of Australian hard woods, but so far have failed. Asphalt—Have laid several streets with tar macadam.	—	—	We are purchasing a new 15-tons Aveling and Porter's road engine with Morrison's scarifier, which I anticipate will greatly economise labour.	I use both qualities, and I am satisfied that hand-broken has a longer life than machine-broken.	—	We do not use salt water.

The material originally positioned here is too large for reproduction in this reissue. A PDF can be downloaded from the web address given on page iv of this book, by clicking on 'Resources Available'.

DETAILS RELATING TO ROAD CONSTRUCTION, TABULATED BY THE AUTHOR, FROM REPLIES RECEIVED FROM SOME LEADING ENGINEERS, EXPERIENCED IN THIS IMPORTANT BRANCH OF CIVIL ENGINEERING, AND TO WHOM THE AUTHOR IS INDEBTED FOR THE VALUABLE INFORMATION (*continued*).

Authority.	With what description of Materials are your Roads Maintained?	What is the Approximate Average Thickness Worn per Annum, and Description of Traffic over same?	What is the Approximate Initial and Annual Cost of the Materials you use?	Do you adopt a Road Scarifier or Hand Labour for Breaking your Roads? State the advantages of the former, if any.	Is your Macadam broken by Hand or Machine? Do you find the Life of the Material reduced by one or the other?	What save, or otherwise, do you consider there is in Scavenging Wood and Asphalt Roads, as compared with Macadam Roads?	Do you use Salt Water for Street Watering? If so, with what advantage, or otherwise?
Francis J. C. May, Brighton	Macadam—Guernsey granite and hand-picked flints. Wood—Australian hard wood and also creosoted Swedish deal.	Macadam wears about 1in. to 1½in. In the business thoroughfares it is very heavy, and includes a five minutes' service of omnibuses, generally heavy laden. On the principal drives continuous carriage traffic. Hard wood wears about ½in. Soft wood about ½in. to ¾in.	Macadam. Repairs about 1s. 6d. per yard. Initial cost varies according to mode of construction. Hard wood. Repairs about 1½d. per yard. Initial cost 17s. 10d. per yard super. Soft wood. Repairs about 3d. per yard. Initial cost, 14s. 8d. per yard super.	We use a scarifier, "Jackson's" patent. I find the saving about 66 per cent. over that of hand labour.	It is mostly machine-broken; life is not thereby reduced to any appreciable extent, but I consider hand-broken material is decidedly the best.	The cost of scavenging macadam roads is about 9d. per yard super per annum. The wood paved roads about 2½d. per yard. The latter roads are all our best business thoroughfares, and are particularly well scavenged; under ordinary conditions the cost should be less.	No.
C. Brown, Wells, Somerset	Macadam—Limestone.	Macadam wears under an inch; traffic comparatively light.	Macadam. 3s. to 4s. per ton in place.	I have tried the road scarifier, but find the cost is very much in excess of hand labour, that I have abandoned its use.	Macadam broken by hand, which is unquestionably more lasting than machine-broken stone.	—	No.
Rees Jones, Aberystwyth	Macadam—North Wales Granite. Wood.	Macadam wears about 1in.; ordinary light traffic.	Macadam. The cost of broken macadam is 5s. 9d. per ton. About 1000 tons used annually.	I have not used a scarifier.	By machine. I consider hand-broken macadam at least 10 per cent. better than machine-broken.	Had no experience with wood or asphalt.	No.
John S. Brodie, Blackpool	Macadam—Broken syenite, limestone, &c. Wood—Pitch pine and Jarrah wood. Asphalt—Tar asphalt, limestone, metalling.	Macadam. ¾in. to 1in. Hard wood. About 1/16in. Soft wood. About 3/16in. Asphalt. ¼in. to ½in.	Macadam. 5s. 6d. per square yard Hard wood. 14s. per square yard Soft wood. 11s. per square yard Asphalt. 6s. 6d. per square yard } Initial cost. I have no present statistics as to annual cost of repairs. Present prices, which are very high.	No.	Chiefly by machine, being mostly bought ready broken.	Very large amount, but no reliable statistics.	—
M. B. Y. Bennett, Southampton ...	Macadam—Guernsey, Penlee, Cherbourg, and German basalt, also flints and local gravel. Wood—Plain deal, creosoted and dipped deal. West Australian hardwood. Asphalt—French Asphalt Co.'s	Macadam worn about ½in. per annum, but not sufficient data to give an exact figure. Hard wood. Karri blocks in narrow street, with heavy traffic, had not worn 1/16in. in two years. Soft wood. Creosoted and pickled deal has deteriorated to a much greater extent in two years. Asphalt. 2in. of compressed asphalt in the High-street was down 28 years without renewal.	Macadam. 10s. to 13s. per ton. Hard wood. About 10s. per yard super, blocks only. Soft wood. About 5s. per yard super, blocks only. Asphalt. 15s. per yard super, laid.	Hand labour employed.	Mostly hand-broken.	Very little, as we make it a point to thoroughly cleanse wood and asphalt paving, and constantly remove the droppings.	No.
R. H. Morgan, Bolton	Macadam—Granite, limestone. Wood—Red deal and beech.	Macadam. 1in., ordinary manufacturing town traffic only. Hard wood. Beech, heavy traffic. Soft wood. Red deal, heavy traffic. This pavement has only been laid a short time, so cannot give thickness worn.	Macadam. Granite, 11s. 3d. per ton; limestone, 6s. 6d. per ton.	Yes. Aveling and Porter. ¼d. per yard.	Machine. No; doubtful.	—	No.
Frank Baker, Middlesboro	Macadam—Whinstone or basalt.	—	—	We have a Jackson patent scarifier.	Machine-broken.	—	No.
Hy. Marks, Carlisle	Macadam—A few miles of macadam. Whinstone (a basaltic rock), 2½in. hand-broken. Wood—Small amount of Karri block paving.	Macadam. Generally about ½in. Chiefly centre of road, ordinary traffic.	—	Aveling and Porter's scarifier. Does three times the work of hand labour at a given cost.	Hand-broken. Considered very much superior.	—	No.
J. C. Butland, Belfast	Macadam—Local whinstone and basalt.	Macadam. Varies according to traffic; no statistics; we use about 50,000 tons per annum.	Macadam. Costs 4s. per ton on streets ready to spread. Original cost of macadam roads on 9in. pitching, 3s. 6d. per square yard; repairs according to traffic.	Roads backed by hand labour.	Both. Hand-broken macadam is best for all purposes.	I think the labour will be greater in the case of wood or asphalt, but less material to remove.	No, but think it would be a saving in volume of water required.
R. M. Gloyne, Eastbourne	Macadam—Mostly quartzite, 1½in. gauge, and flints hand-broken.	Light traffic generally; average thickness not noted.	Macadam. Quartzite, 14s., delivered in depôt, per ton. Flints, 10s. per cubic yard, delivered on roads.	Scarifier, Morrison's, about ¼d. per square yard to 1d. in actual disturbance, but allowance must also be made for the greater regularity of work. When hand labour is used, the hollows are picked up to as great a depth as the humps.	Quartzite machine-broken. Flints hand-broken. Neither material is broken except as stated, so no comparison is available.	—	No. The question has often been raised, and I have reported on a scheme, but the disadvantages where trees are planted in the streets, and objections taken by private owners of carriages, in addition to probable cost of an installation of pumps, reservoirs, pipes, &c., are thought to considerably outweigh the advantages.
John F. Smithe, Tynemouth... ...	Macadam—Principally Northumberland whinstone.	—	Macadam. Initial cost of road, including 9in. formation of hand-pressed rubble and 9in. of metal, 6s. per square yard. Annual cost of maintenance varies from £250 per mile to £400 per mile.	Have no road scarifier. Road first loosened with pins on one of big wheels of roller, and then breaking finished with hand labour.	Both. Northumbrian whinstone breaks badly with Blake Marsden machine. Prefer hand-broken metal.	—	Yes. In dry weather great advantage, as a crust is formed upon the surface of the road which binds the dust. In humid weather—not rainy—a nuisance. The salt "comes" again and makes the roads sticky and greasy, and if such weather long continues, the roads become so very sticky that the traffic lifts the metal.
Philip Murch, Portsmouth	Macadam—Broken Guernsey granite. Asphalt—Limmer asphalt.	Macadam. The average over the whole town would be less than 1in. per annum—in the main streets, with heavy traffic, it would be about 1½in. Asphalt. About ½in. in ten years; omnibus and trade traffic.	Macadam. Broken Guernsey granite 12s. per ton on quay; laid and rolled in, average cost, 2½in. thick, would be 2s. per yard super. Asphalt. 14s. per yard super. No wood paving.	Yes, one by Aveling and Porter, Rochester, Kent; also hand labour. The saving is, I estimate, 33 per cent. in favour of scarifier.	The Guernsey granite is broken by machinery at the quarries.	With regard to our asphalt roads I do not think there is a saving. A man has to constantly attend to the surface, which must be kept very clean otherwise it becomes dangerous to horses, especially in damp weather.	Yes, over a small area of the Borough with this advantage, that one load of salt water is equal to nearly three loads of fresh in laying dust.

The material originally positioned here is too large for reproduction in this reissue. A PDF can be downloaded from the web address given on page iv of this book, by clicking on 'Resources Available'.

Printed in the United States
By Bookmasters